7

Springer Series on Chemical Sensors and Biosensors

Methods and Applications

Series Editor: G. Urban

For further volumes:
http://www.springer.com/series/5346

Springer Series on Chemical Sensors and Biosensors

Series Editor: G. Urban

Recently Published and Forthcoming Volumes

Optical Guided-wave Chemical and Biosensors II
Volume Editors: Zourob M., Lakhtakia A.
Vol. 8, 2010

Optical Guided-wave Chemical and Biosensors I
Volume Editors: Zourob M., Lakhtakia A.
Vol. 7, 2010

Hydrogel Sensors and Actuators
Volume Editors: Gerlach G., Arndt K. -F.
Vol. 6, 2009

Piezoelectric Sensors
Volume Editors: Steinem C., Janshoff A.
Vol. 5, 2006

Surface Plasmon Resonance Based Sensors
Volume Editor: Homola J.
Vol. 4, 2006

Frontiers in Chemical Sensors
Novel Principles and Techniques
Volume Editors: Orellana G., Moreno-Bondi M. C.
Vol. 3, 2005

Ultrathin Electrochemical Chemo- and Biosensors
Technology and Performance
Volume Editor: Mirsky V. M.
Vol. 2, 2004

Optical Sensors
Industrial, Environmental and Diagnostic Applications
Volume Editors:
Narayanaswamy R., Wolfbeis O. S.
Vol. 1, 2003

Optical Guided-wave Chemical and Biosensors I

Volume Editors: Mohammed Zourob • Akhlesh Lakhtakia

With contributions by

N. R. Armstrong · A. G. Brolo · D. P. Campbell · Q. J. Cheng
R. Gordon · C. Hoffmann · D. Kim · M. J. Linman · S. B. Mendes
S. Mittler · S. S. Saavedra · K. E. Sapsford · K. Schmitt · D. Sinton

Springer

Chemical sensors and biosensors are becoming more and more indispensable tools in life science, medicine, chemistry and biotechnology. The series covers exciting sensor-related aspects of chemistry, biochemistry, thin film and interface techniques, physics, including opto-electronics, measurement sciences and signal processing. The single volumes of the series focus on selected topics and will be edited by selected volume editors. The Springer Series on Chemical Sensors and Biosensors aims to publish state-of-the-art articles that can serve as invaluable tools for both practitioners and researchers active in this highly interdisciplinary field. The carefully edited collection of papers in each volume will give continuous inspiration for new research and will point to existing new trends and brand new applications.

ISSN 1612-7617
ISBN 978-3-642-26296-8 e-ISBN 978-3-540-88242-8
DOI 10.1007/978-3-540-88242-8
Springer Heidelberg Dordrecht London New York

Softcover reprint of the hardcover 1st edition 2009

Cover design: WMXDesign GmbH, Heidelberg

Printed on acid-free paper

Springer is part of Springer Science+Business Media (www.springer.com)

Series Editor

Prof. Dr. Gerald Urban

IMTEK - Laboratory for Sensors
Institute for Microsystems Engineering
Albert-Ludwigs-University
Georges-Köhler-Allee 103
79110 Freiburg
Germany
urban@imtek.de

Volume Editors

Dr. Mohammed Zourob

GDG ENVIRONNEMENT LTÉE
430, rue Saint-Laurent
Trois-Rivières (Quebec)
G8T 6H3 Canada
mohammed.zourob@gdg.ca

INRS Énergie, Matériaux et Télécommunications
1650, boul. Lionel-Boulet
Varennes, Québec
J3X 1S2 Canada
zourob@emt.inrs.ca

Prof. Dr. Akhlesh Lakhtakia

Pennsylvania State University
Dept. Engineering Science & Mechanics
University Park PA 16802
USA
axl4@psu.edu

Springer Series on Chemical Sensors and Biosensors Also Available Electronically

For all customers who have a standing order to Springer Series on Chemical Sensors and Biosensors, we offer the electronic version via SpringerLink free of charge. Please contact your librarian who can receive a password or free access to the full articles by registering at:

springerlink.com

If you do not have a subscription, you can still view the tables of contents of the volumes and the abstract of each article on SpringerLink. Just click on "Online version available" on the series homepage (www.springer.com/series/5346).

Preface

Optical sensing techniques based on the modification of the refractive index because of either the incursion of a chemical species (analyte) or interactions between two different types of chemical species, one of which is the analyte and the other is the ligand, have had a long history that goes back to beginning of the nineteenth century if not earlier. More recently, fluorescence resulting from an appropriate labeling/modification of the ligand–analyte interaction has been used for optical sensing. Applications of these techniques are commonplace in industry to ascertain the density of a manufactured material in solution; in biomedical labs to detect the presence of a toxin or biological analytes in a fluid; and so on.

Two recent developments have further spurred research on the sensing of chemicals and analytes of biological importance. First, in the aftermath of the horrific events that occurred on September 11, 2001, a major nightmare of homeland-security planners and providers is the deliberate introduction of toxins and pathogens – which include pesticides (e.g., atrazine and 2,4-dichlorophenol) and bacteria (such as *Vibrio cholerae* and *Salmonella paratyphi*) – in a nation's water resources and urban water-distribution systems. Rapidly acting toxins and pathogens may be released in ponds, lakes, and rivers by enemy troops in war-zones to disable the soldiers fighting them. The same strategy may also be employed by guerrilla fighters sneaking behind the forward positions of regular armed forces. Thus, multianalyte sensing systems with remote monitoring capabilities, high sensitivity, and low incidence of false results are urgently needed to quickly assess both internal and external threats implemented by saboteurs.

Second, our air and water are increasingly more polluted from relentless industrialization in many rapidly developing countries and the conversion of farming from a family-based enterprise to agribusiness throughout the world. Water pollution is already a serious global problem, as the contents of wastewaters have become very complex. Chemical mutagens and carcinogens derived from industrial waste, pesticides, and urban sewage are known to cause metabolic damage in living organisms. For example, endocrine-disrupting compounds are able to either mimic or counter antagonize the effects of hormones such as estrogens and androgens. When present in the environment, endocrine-disrupting compounds engender

Contents of Volume I

Part I
Planar-Waveguide Sensors

Abbreviations

AWACSS	Automated water analyzer computer supported system
CMOS	Complementary metal oxide semiconductor
DCC	Dicyclohexylcarbodiimide
DNA	Deoxyribonucleic acid
EDC	1-ethyl-3-(3-dimethylaminopropyl) carbodiimide hydrochloride
ELISA	Enzyme-linked immunosorbant assay
IgG	Immunoglobulin G
IOW	Integrated optical waveguide
IR	Infrared
IRE	Internal reflection elements
ITO	Indium tin oxide
LED	Light emitting diode
LOD	Limit of detection
MCLW	Metal clad leaky waveguides
MEF	Metal-enhanced fluorescence
NHS	*N*-hydroxysuccinimde
Ni-NTA	Nickel-nitriloacetic acid
NRL	Naval Research Laboratory
OW	Optical waveguide
PDMS	Polydimethylsiloxane
RNA	Ribonucleic acid
SEB	Staphylococcal enterotoxin B
TIR	Total internal reflection
TIRF	Total internal reflection fluorescence
UV	Ultra Violet

1 Introduction

The field of optical-based chemical and biological sensing is vast, encompassing many potential applications including medical, food and water safety, environmental monitoring, national security, and military, as well as understanding fundamental biological processes [1, 2]. Sensing technologies based on total internal reflection (TIR) share an inherent surface-specific sensitivity and come in a variety of different formats demonstrating the unique versatility of this underlying phenomenon [2]. Transduction principles include spectroscopic (infrared, Raman), fluorescence (total internal reflection fluorescence (TIRF)), and refractometry (surface plasmon resonance, interferometric) [2, 3]. Each transduction mechanism has inherent advantages and disadvantages, with the exact choice dependent on a number of factors, including the chosen sensing mechanism and the target analyte.

TIR-based techniques that use fluorescence transduction (TIRF) have been used for the majority of the applications discussed above. These platforms, while not currently the gold standard for measurement of many target analytes, do offer advantages over current technologies, such as cell culture, chromatography, and the enzyme-linked immunosorbant assay (ELISA). Advantages include the ability to perform faster, more sensitive, multiple analyte, and real-time measurements.

2 Principles of TIR

2.1 Basic Principles

When incident light undergoes total reflection and no refraction at an interface between two media of different refractive indices (n_1 and n_2), TIR occurs; see Fig. 1. The TIR phenomenon requires a couple of conditions: (1) the incident light must be traveling in the high-refractive-index medium (n_1, where $n_1 > n_2$) and (2) the angle of incidence (θ) must be greater than the critical angle (θ_c), which is defined by the media interface, as follows:

$$\theta_c = \sin^{-1}(n_1/n_2) \tag{1}$$

TIR results in the generation of an electromagnetic evanescent wave at the point of reflection, decaying exponentially into the lower refractive index media. The depth at which the intensity of either the electric or magnetic component of the evanescent wave drops to $1/e$ of its original value (Equation 2; where $n_{21} = n_2/n_1$) is defined as the depth of penetration (d_p). The extent of d_p into the lower refractive index medium is dependent on a number of physical properties of the system, including the wavelength of the incident light (λ), the angle the incident light is coupled into the waveguide (θ), and the refractive indices at the dielectric interface (n_1 and n_2). Typical d_p values are quoted anywhere between 100 and 300 nm and demonstrate the inherent surface-specific nature of the evanescent wave.

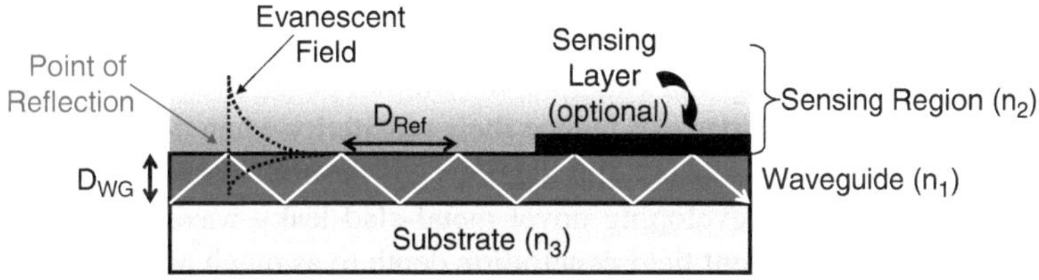

Fig. 1 Schematic illustration of the principle of TIR. Incident light traveling in the high refractive index medium (n_1) undergoes TIR at the interface with the lower refractive index medium (n_2), generating an evanescent wave at the point of reflection

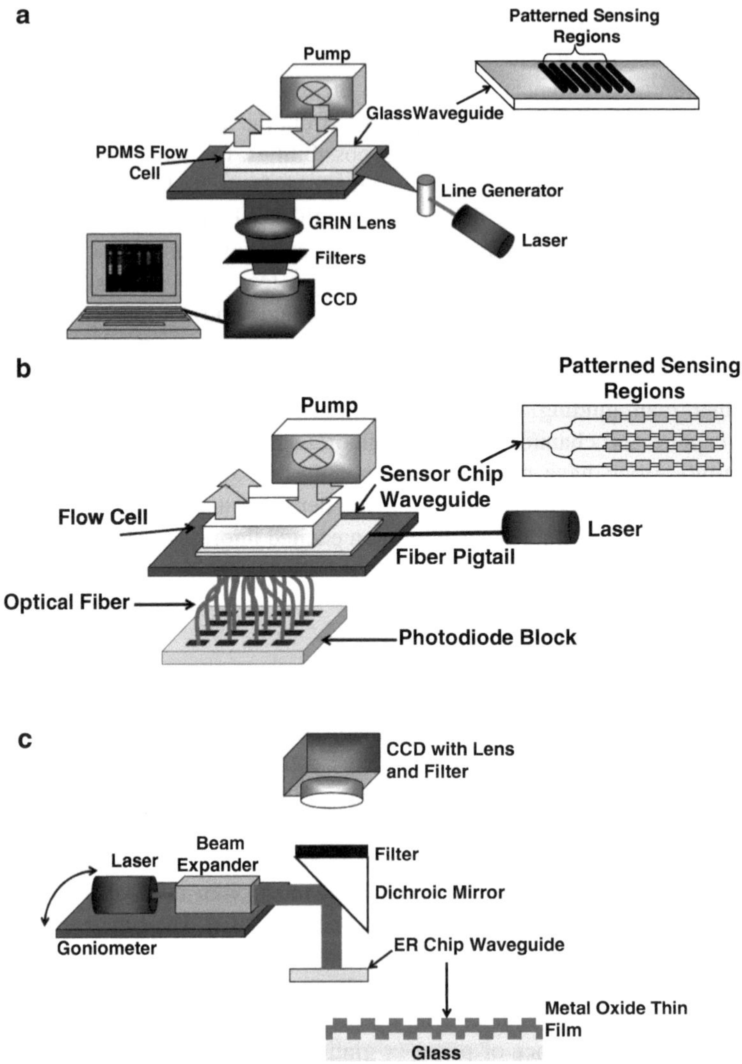

Fig. 2 Schematics illustrating various TIRF-based sensing platforms. (**a**) The NRL Array Biosensor, modified from [28]. (**b**) The AWACSS platform, modified from [29]. (**c**) The Evanescent Resonator (ER) platform modified from [13]

3 Chemical and Biological Sensing Elements

Chemical sensing is defined as the measured response of a chemical compound interacting with a specific analyte and includes the common use of fluorescent devices that monitor a variety of environmentally sensitive dyes. Biological

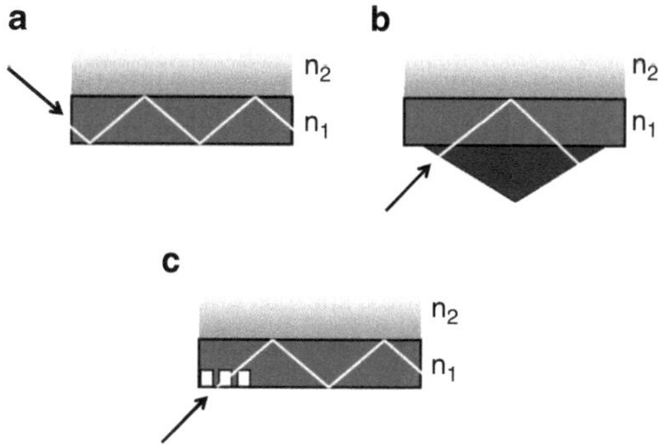

Fig. 3 Schematic illustrating methods of coupling incident light into the waveguide. (**a**) Simple end firing. (**b**) Prism coupling, typically requires the presence of index matching fluid. (**c**) Grating coupling

sensing, in contrast, refers to platforms that incorporate a sensing element of biologically derived origin in intimate contact with the transducer and these usually encompass antibody, nucleic acid, and numerous other biological materials.

Environmentally sensitive fluorophores represent the bulk of chemical-based sensing platforms that use TIRF transduction, with MacCraith and coworkers at the forefront of this field [25, 32]. Many fluorophores are photophysically "sensitive" to changes in their environment, such as pH, ionic strength or type (K^+, Ca^{2+}, Na^+, Cl^- etc), oxygen saturation, salvation, and polarity [33]. Mechanisms via which fluorophores "sense" their environment include: (1) fluorophore emission is collisionally quenched by increasing concentration of a particular ion, (2) the presence of a target species can cause a spectral shift in the excitation/emission profiles of the fluorophore, and (3) the fluorophore may display target-concentration-dependent emission intensity (but no spectral shift) [33]. To date, only a limited number of environmentally sensitive fluorophores have been applied to TIRF-based chemical sensing, with pH and oxygen sensing representing the bulk of the research.

In contrast, biological sensing has enjoyed a great deal of research activity. Most of the biological TIRF sensing platforms fall under the definition of affinity-based sensors, with antibodies and nucleic acids representing the more popular choice of sensing element [34]. Unlike direct chemical sensors, most biological sensors require the addition of a fluorescent component following target recognition, which increases the complexity of the sensing event. However, unlike the chemical-based sensors, the number of potential targets measured using biological sensing elements is vast. Antibodies, for example, have been generated to species as varied as simple organic compounds to whole bacterial cells [35] and are readily commercially available. Likewise, specific hybridization with DNA or RNA can be used to measure DNA/RNA containing targets, such as bacteria and viruses. Nucleic acid hybridization also benefits from the ability to perform PCR amplification of the target DNA/RNA, improving sensitivity and surface regeneration after use, leading to reusable sensors [36–40]. Other promising biological sensing elements that have already demonstrated TIRF applicability include aptamers, receptors, carbohydrates, and peptides [41–46].

by [28]). Using antibody-based sandwich or competitive formats, the assay performance in a number of complex matrices, including clinical, environmental, and food matrices, has been evaluated to determine matrix effects on the overall sensitivity and reliability of this platform ([28] and references therein). In many instances, the matrix was found to influence the sensitivity of the assay, highlighting the need for matrix calibration of the sensing platform response prior to characterization of unknown samples [45, 79, 80]. The NRL platform has also been used to study the real-time binding kinetics of both antibody–antigen (specific) [81], carbohydrate–protein (semi-selective) [44] and protein-surface (nonspecific) [82] interactions.

DNA hybridization and increasingly RNA studies have been investigated by a number of researchers using TIRF-based transduction techniques. Budach et al. [36] and Duveneck et al. [31] demonstrated oligonucleotide hybridization assays on Ta_2O_5 IOW surfaces, with LODs in the fM range. Budach and coworkers later modified their IRE waveguide to produce the ER chips described earlier, which have been applied to gene expression/RNA detection [11–13]. Schuderer and coworkers demonstrated pM detection using bulk glass waveguides [40]. Real-time monitoring of DNA hybridization assays have also been successfully demonstrated using bulk glass [83] and polystyrene [84] waveguides via TIRF, with LODs in the pM range. Real time kinetic studies allow the researcher to study the shape of the hybridization curve, which in turn can be used to detect the presence of mismatched pairs [84].

Recently, researchers have been investigating the potential application of alternative, often semi-selective, capture species such as purified carbohydrates, gangliosides, and antimicrobial peptides. A number of these have been developed using the NRL Array Biosensor and illustrate the exciting possibility of pathogen detection based on fingerprint/pattern recognition, with the potential to detect "unexpected" targets [41–46].

5 Multiplexing TIRF – Array Sensing

There are two major advantages to the use of planar waveguides for sensing technologies. The first is the ability to define patterns of immobilized sensing elements on a single surface, thus allowing for the possibility of multiple analyte detection on a single substrate through spatial separation. The second concerns the ease of integration into microfluidic platforms, allowing for the possibility of semi-to-fully automated assay control.

While Malins and coworkers demonstrated multianalyte, chemical sensing by patterning sol-gel thin-film-immobilized fluorescent dyes, the majority of multiplexed TIRF has resulted from the use of biological-based sensing elements [32]. Techniques available for patterning biological molecules onto planar waveguide surfaces generally involve either photopatterning, printing or physically isolated deposition. Photopatterning or photolithography, as the name suggests, involves coating a surface with a light-sensitive material, in the form of a thin-film monolayer, laminate, photoresist or gel, and could be considered a form of noncontact printing [81, 85–87]. Subsequent use of a mask and light of the appropriate wavelength (typically UV), or the technique of laser writing creates regions of the waveguide either reactive to or resistant to biomolecule adsorption. Such techniques have the ability to generate very high-density arrays of sensing features on the surface of the waveguide [90, 92]. While very effective, the use of photopatterning for immobilization of multiple different capture biomolecules can be more complicated, involving multiple steps and multiple chemistries [92–94].

Printing techniques can be broadly grouped into contact and noncontact methods, reviewed by Barbulovic-Nad and coworkers [95], and include pin-printing contact arrayers, inkjet printing, and microstamping using elastomeric stamps. These printing methods are very popular and effective techniques for producing high-density multiple capture arrays on planar surfaces [36, 95–102]. Much research effort has been invested in determining the optimal printing conditions to produce uniform, dense arrays in a reproducible fashion [70].

As an alternative to printing, physically isolated patterning has been demonstrated in various formats by a number of researchers [40, 103–108]. Examples include the use of PDMS flow cells, silicon wafer stencils, and 96-well plate formats [28, 108, 109]. Figure 4 demonstrates patterning using a vinyl template/stencil, cut using a laser cutter, and subsequently attached to an avidin-functionalized glass surface. The resulting 1 μL wells were filled using a laboratory pipette with biotin-functionalized proteins before direct immunoassays were performed with the aid of a second vinyl template. While more suited to the production of lower density arrays, compared to those achievable via printing, these techniques

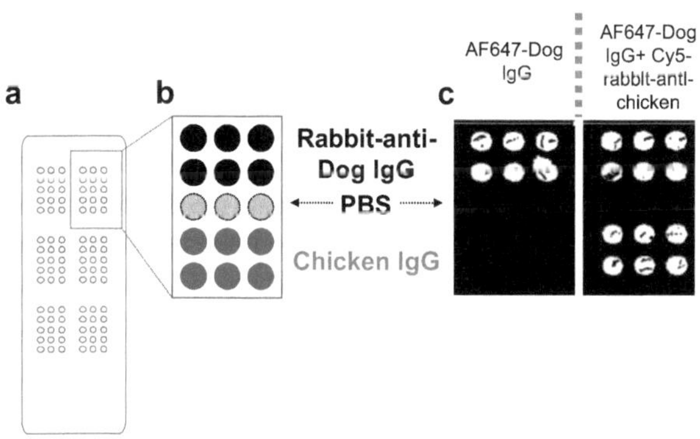

Fig. 4 Patterning waveguide surfaces. (**a**) Vinyl template/stencil is cut using a laser cutter before being attached to the waveguide surface. Pattern gives a total of 90 × 1 μL wells, arranged in six regions, which are patterned as illustrated in (**b**) with rabbit-anti-dog IgG, PBS, and chicken IgG. (**c**) Image recorded using a Genepix 4000 slide scanner of a patterned glass surface following a direct immunoassay of two separate regions of the same slide. One region was exposed to AlexaFluor647-dog IgG only, while the second region was exposed to both AlexaFluor647-dog IgG plus Cy5-rabbit-anti-chicken IgG (previously unpublished data from author K. Sapsford)

References

1. Bally M, Halter M, Voros J et al (2006) Optical microarray biosensing techniques. Surf Interface Anal 38:1442–1458
2. Ince R, Narayanaswamy R (2006) Analysis of the performance of interferometry, surface plasmon resonance and luminescence as biosensors and chemosensors. Anal Chim Acta 569:1–20
3. Potyrailo RA, Hobbs SE, Hieftje GM (1998) Optical waveguide sensors in analytical chemistry: todays instrumentation, applications and trends for future development. Fresenius J Anal Chem 362:349–373
4. Bernini R, Cennamo N, Minardo A (2006) Planar waveguides for fluorescence-based biosensing: optimization and analysis. IEEE Sens J 6:1218–1226
5. Schmidt H, Hawkins AR (2008) Optofluidic waveguides: I. Concepts and implementations. Microfluid Nanofluid 4:17–32
6. Schmitt K, Oehse K, Sulz G et al (2008) Evanescent field sensors based on tantalum pentoxide waveguides – a review. Sensors 8:711–738
7. Lehr HP, Brandenburg A, Sulz G (2003) Modeling and experimental verification of the performance of TIRF-sensing systems for oligonucleotide microarrays based on bulk and integrated optical planar waveguides. Sensors Actuators B Chem 92:303–314
8. Zourob M, Mohr S, Treves-Brown BJ et al (2005) An integrated metal clad leaky waveguide sensor for detection of bacteria. Anal Chem 77:232–242
9. Zourob M, Hawkes JJ, Coakley WT et al (2005) Optical leaky waveguide sensor for detection of bacteria with ultrasound attractor force. Anal Chem 77:6163–6168
10. Zourob M, Simonian A, Wild J et al (2007) Optical leaky waveguide biosensors for the detection of organophosphorus pesticides. The Analyst 132:114–120
11. Beuvink I, Kolb FA, Budach W et al (2007) A novel microarray approach reveals new tissue-specific signatures of known and predicted mammalian microRNAs. Nucleic Acids Res 35:e52
12. Budach W, Neuschafer D, Wanke C et al (2003) Generation of transducers for fluorescence-based microarrays with enhanced sensitivity and their application for gene expression profiling. Anal Chem 75:2571–2577
13. Neuschafer D, Budach W, Wanke C et al (2003) Evanescent resonator chips: a universal platform with superior sensitivity for fluorescence-based microarrays. Biosens Bioelectron 18:489–497
14. Chen Y, Munechika K, Ginger DS (2007) Dependence of fluorescence intensity on the spectral overlap between fluorophores and plasmon resonant single silver nanoparticles. Nano Lett 7:690–696
15. Ganesh N, Zhang W, Mathias PC et al (2007) Enhanced fluorescence emission from quantum dots on a photonic crystal surface. Nat Nanotechnol 2:515–520
16. Konopsky VN, Alieva EV (2007) Photonic crystal surface waves for optical biosensors. Anal Chem 79:4729–4735
17. Lakowicz JR (2006) Plasmonics in biology and plasmon-controlled fluorescence. Plasmonics 1:5–33
18. Mathias PC, Ganesh N, Chan LL et al (2007) Combined enhanced fluorescence and label-free biomolecular detection with a photonic crystal surface. Appl Opt 46:2351–2360
19. Moores A, Goettmann F (2006) The plasmon band in noble metal nanoparticles: An introduction to theory and applications. New J Chem 30:1121–1132
20. Stranik O, Nooney R, McDonagh C et al (2007) Optimization of nanoparticle size for plasmonic enhancement of fluorescence. Plasmonics 2:15–22
21. Zhang YQ, Wang JX, Ji ZY et al (2007) Solid-state fluorescence enhancement of organic dyes by photonic crystals. J Mater Chem 17:90–94

22. Aslan K, Holley P, Geddes CD (2006) Microwave-accelerated metal-enhanced fluorescence (MAMEF) with silver colloids in 96-well plates: Application to ultra fast and sensitive immunoassays, high throughput screening and drug discovery. J Immunol Methods 312:137–147
23. Matveeva E, Gryczynski Z, Malicka J et al (2004) Metal-enhanced fluorescence immunoassays using total internal reflection and silver-coated surfaces. Anal Biochem 334:303–311
24. Blue R, Kent N, Polerecky L (2005) Platform for enhanced detection efficiency in luminescent-based sensors. Electron Lett 41:682–684
25. Burke CS, McGaughey O, Sabattie J-M et al (2005) Development of an integrated optic sensor using a novel, generic platform. The Analyst 130:41–45
26. Holthoff WG, Tehan EC, Bukowski RM et al (2005) Radioluminescent light source for the development of optical sensors arrays. Anal Chem 77:718–723
27. Schultz E, Galland R, Du Bouetiez D et al (2008) A novel fluorescence-based array biosensor: principle and application to DNA hybridization assays. Biosens Bioelectron 23:987–994
28. Ligler FS, Sapsford KE, Golden JP et al (2007) The array biosensor: portable, automated systems. Anal Sci 23:5–10
29. Hua P, Hole JP, Wilkinson JS et al (2005) Integrated optical fluorescence multisensor for water pollution. Opt Express 13:1124–1130
30. Golden JP, Ligler FS (2002) A comparison of imaging methods for use in an array biosensor. Biosens Bioelectron 17:719–725
31. Duveneck GL, Abel AP, Bopp MA (2002) Planar waveguides for ultra-high sensitivity of the analysis of nucleic acids. Anal Chim Acta 469:49–61
32. Malins C, Niggemann M, MacCraith BD (2000) Multi-analyte optical chemical sensor employing a plastic substrate. Meas Sci Technol 11:1105–1110
33. Lakowicz JR (2006) Principles of fluorescence spectroscopy, 3rd edn. Springer, New York
34. Marquette CA, Blum LJ (2006) State of the art and recent advances in immunoanalytical systems. Biosens Bioelectron 21:1424–1433
35. Moore P, Clayton J (2003) To affinity and beyond. Nature 426:725–731
36. Budach W, Abel AP, Bruno AP et al (1999) Planar waveguides as high performance sensing platforms for fluorescence-based multiplexed oligonucleotide hybridization assays. Anal Chem 71:3347–3355
37. Duveneck GL, Pawlak M, Neuschafer D et al (1997) Novel bioaffinity sensors for trace analysis based on luminescence excitation by planar waveguides. Sensors Actuators B Chem 38:88–95
38. Lim DV, Simpson JM, Kearns EA et al (2005) Current and developing technologies for monitoring agents of bioterrorism and biowarfare. Clin Microbiol Rev 18:583–607
39. Schaferling M, Nagl S (2006) Optical technologies for the read out and quality control of DNA and protein microarrays. Anal Bioanal Chem 385:500–517
40. Schuderer J, Akkoyun A, Brandenburg A et al (2000) Development of a multichannel fluorescence affinity sensor system. Anal Chem 72:3942–3948
41. Kulagina NV, Lassman ME, Ligler FS et al (2005) Antimicrobial peptides for detection of bacteria in biosensor assays. Anal Chem 77:6504–6508
42. Kulagina NV, Shaffer KM, Anderson GP et al (2006) Antimicrobial peptide-based array for Escherichia coli and Salmonella screening. Anal Chim Acta 575:9–15
43. Kulagina NV, Shaffer KM, Ligler FS et al (2007) Antimicrobial peptides as new recognition molecules for screening challenging species. Sensors Actuators B Chem 121:150–157
44. Ngundi MM, Taitt CR, Ligler FS (2006) Simultaneous determination of kinetic parameters for the binding of cholera toxin to immobilized sialic acid and monoclonal antibody using an array biosensor. Biosens Bioelectron 22:124–130
45. Ngundi MM, Qadri SA, Wallace EV et al (2006) Detection of deoxynivalenol in foods and indoor air using an array biosensor. Environ Sci Technol 40:2352–2356
46. Rowe-Taitt CA, Cras JJ, Patterson CH et al (2000) A ganglioside-based assay for cholera toxin using an array biosensor. Anal Biochem 281:123–133

47. Ngundi MM, Taitt CR, McMurry SA et al (2006) Detection of bacterial toxins with monosaccharide arrays. Biosens Bioelectron 21:1195–1201
48. Peterson AW, Heaton RJ, Georgiadis RM (2001) The effect of surface probe density on DNA hybridization. Nucleic Acids Res 29:5163–5168
49. Zhi ZL, Powell AK, Turnbull JE (2006) Fabrication of carbohydrate microarrays on gold surfaces: direct attachment of noderivatized oligosaccharides to hydrazide-modified self-assembled monolayers. Anal Chem 78:4786–4793
50. Rusmini F, Zhong Z, Feijen J (2007) Protein immobilization strategies for protein biochips. Biomacromol 8:1775–1789
51. Hermanson GT (1996) Bioconjugate techniques. Academic, San Diego
52. Bilitewski U (2006) Protein-sensing assay formats and devices. Anal Chim Acta 568:232–247
53. Pierce. (2008) Avidin biotin products. http://www.piercenet.com/Objects/View.cfm?Type=Page&ID=DDDF4C6A-DF1E-40D5-A01D-870EA98CBF19. Accessed 09 Sept, 2008
54. Bonroy K, Frederix F, Reekmans G, Dewolf E, De Palma R, Borghs G, Declerck P, Goddeeris B (2006) Comparison of random and oriented immobilization of antibody fragments on mixed self-assembled monolayers. J Immunol Methods 312:167–181
55. Johnson CP, Jensen IE, Prakasam A et al (2003) Engineered protein A for the orientational control of immobilized proteins. Bioconjug Chem 14:974–978
56. McLean MA, Stayton PS, Sligar SG (1993) Engineering protein orientation at surfaces to control macromolecular recognition events. Anal Chem 65:2676–2678
57. Shen Z, Tan H, Zhang Y et al (2008) Engineering peptide linkers for scFv immunosensors. Anal Chem 80:1910–1917
58. Valsesia A, Colpo P, Mannelli I et al (2008) Use of nanopatterned surfaces to enhance immunoreaction efficiency. Anal Chem 80:1418–1424
59. Vigmond SJ, Iwakura M, Mizutani F et al (1994) Site-specific immobilization of molecularly engineered dihydrofolate-reductase to gold surfaces. Langmuir 10:2860–2862
60. Vijayendran RA, Leckband DE (2001) A quantitative assessment of heterogeneity for surface-immobilized proteins. Anal Chem 73:471–480
61. Vikholmn I, Albers WM (1998) Oriented immobilization of antibodies for immunosensing. Langmuir 14:3865–3872
62. Beaucage SL (2001) Strategies in the preparation of DNA oligonucleotide arrays for diagnostic applications. Curr Med Chem 8:1213–1244
63. Dufva M (2005) Fabrication of high quality microarrays. Biomol Eng 22:173–184
64. Wellman AD, Sepaniak MJ (2006) Magnetically-assisted transport evanescent field fluoroimmunoassay. Anal Chem 78:4450–4456
65. Wellman AD, Sepaniak MJ (2007) Multiplexed, waveguide approach to magnetically assisted transport evanescent field fluoroassays. Anal Chem 79:6622–6628
66. Loete F, Vuillemin B, Oltra R et al (2006) Application of total internal reflection fluorescence microscopy for studying pH changes in an occluded electrochemical cell: development of a waveguide sensor. Electrochem Commun 8:1016–1020
67. Ichinose J, Sako Y (2004) Single-molecule measurement in living cells. Trends Anal Chem 23:587–594
68. Mashanov GI, Tacon D, Knight AE et al (2003) Visualizing single molecules inside living cells using total internal reflection fluorescence microscopy. Methods 29:142–152
69. Webb SED, Needham SR, Roberts SK et al (2006) Multidimensional single-molecule imaging in live cells using total-internal-reflection fluorescence microscopy. Opt Lett 31:2157–2159
70. Moran-Mirabal JM, Edel JB, Meyer GD et al (2005) Micrometer-sized supported lipid bilayer arrays for bacterial toxin binding studies through total internal reflection fluorescence microscopy. Biophys J 89:296–305
71. Hoshino A, Fujioka K, Manabe N et al (2005) Simultaneous multicolor detection system of the single-molecular microbial antigen with total internal reflection fluorescence microscopy. Microbiol Immunol 49:461–470

72. Rodriguez-Mozaz S, Reder S, Lopez de Alda MJ et al (2004) Simultaneous multi-analyte determination of estrone, isoproturon, and atrazine in natureal waters by the RIver ANAlyser (RIANA), an optical immunosensor. Biosens Bioelectron 19:633–640
73. Rodriguez Mozaz S, Lopez de Alda MJ, Barcelo D (2006) Biosensors as useful tools for environmental analysis and monitoring. Anal Bioanal Chem 386:1025–1041
74. Tschmelak J, Proll G, Gauglitz G (2004) Verification of performance with the automated direct optical TIRF immunosensor (River analyser) in single and multi-analyte assays with real water samples. Biosens Bioelectron 20:743–752
75. Tschmelak JG, Proll J, Riedt J et al (2005) Part I: project objectives, basic technology, immunoassay development, software design and networking. Part II: intelligent, remote controlled, cost effective, on-line, water monitoring measurement system. Biosens Bioelectron 20:1499–1519
76. Tschmelak J, Kappel N, Gauglitz G (2005) TIRF-based biosensor for sensitive detection of progesterone in milk based on ultra-sensitive progesterone detection in water. Anal Bioanal Chem 382:1895–1903
77. Tschmelak J, Kumpf M, Kappel N et al (2006) Total internal reflection fluorescence (TIRF) biosensor for environmental monitoring of testosterone with commercially available immunochemistry: antibody characterization, assay development and real sample measurements. Talanta 69:343–350
78. Engstrom HA, Andersson PO, Ohlson S (2006) A label-free continuous total-internal-reflection-fluorescence-based immunosensor. Anal Biochem 357:159–166
79. Sapsford KE, Rasooly A, Taitt CR et al (2004) Detection of Campylobacter and Shigella species in food samples using an array biosensor. Anal Chem 76:433–440
80. Sapsford KE, Taitt CR, Loo N et al (2005) Biosensor detection of botulinum toxoid A and staphylococcal enterotoxin B in food. Appl Environ Microbiol 71:5590–5592
81. Sapsford KE, Liron Z, Shubin YS et al (2001) Kinetics of antigen binding to arrays of antibodies in different sized spots. Anal Chem 73:5518–5524
82. Sapsford KE, Ligler FS (2004) Real-time analysis of protein adsorption to a variety of thin films. Biosens Bioelectron 19:1045–1055
83. Lehr HP, Reimann M, Brandenburg A et al (2003) Real-time detection of nucleic acid interactions by total internal reflection fluorescence. Anal Chem 75:2414–2420
84. Tolley SE, Wang HK, Smith RS et al (2003) Single-chain polymorphism analysis in long QT syndrome using planar waveguide fluorescent biosensors. Anal Biochem 315:223–237
85. Bhatia SK, Hickman JJ, Ligler FS (1992) New approach to producing patterned biomolecular assemblies. J Am Chem Soc 114:4432–4433
86. Bhatia SK, Teixeira JL, Anderson M et al (1993) Fabrication of surfaces resistant to protein adsorption and application to 2-dimensional protein patterning. Anal Biochem 208:197–205
87. Liu XH, Wang HK, Herron JN et al (2000) Photopatterning of antibodies on biosensors. Bioconj Chem 11:755–761
88. Arenkov P, Kukhtin A, Gemmell A et al (2000) Protein microchips: use for immunoassay and enzymatic reactions. Anal Biochem 278:123–131
89. Guschin D, Yershov G, Zaslavsky A et al (1997) Manual manufacturing of oligonucleotide, DNA, and protein microchips. Anal Biochem 250:203–211
90. LaFratta CN, Walt DR (2008) Very high density sensing arrays. Chem Rev 108:614–637
91. Wadkins RM, Golden JP, Pritsiolas LM et al (1998) Detection of multiple toxic against using a planar array immunoassay. Biosens Bioelectron 13:407–415
92. Sassolas A, Leca-Bouvier BD, Blum LJ (2008) DNA biosensors and microarrays. Chem Rev 108:109–139
93. Blawas AS, Oliver TF, Pirrung MC et al (1998) Step-and-repeat photopatterning of protein features using caged/biotin-BSA: characterization and resolution. Langmuir 14:4243–4250
94. Conrad DW, Golightley SK, Bart JC (1998) Photoactivatable o-nitrobenzyl polyethylene glycol-silane for the production of patterned biomolecular arrays. US Patent 5,773,308

95. Barbulovic-Nad I, Lucente M, Sun Y et al (2006) Bio-microarray fabrication techniques – a review. Crit Rev Biotechnol 26:237–259
96. Delehanty JB, Ligler FS (2002) A microarray immunoassay for simultaneous detection of proteins and bacteria. Anal Chem 74:5681–5687
97. Delehanty JB, Ligler FS (2003) Method for printing functional protein microarrays. BioTechniques 34:380–385
98. Francois P, Charbonnier Y, Jacquet J et al (2006) Rapid bacterial identification using evanescent-waveguide oligonucleotide microarray classification. J Microbiol Methods 65:390–403
99. Ito Y (2006) Photoimmobilization for microarrays. Biotechnol Prog 22:924–932
100. Renault JP, Bernard A, Bietsch A et al (2003) Fabricating arrays of single protein molecules on glass using microcontact printing. J Phys Chem B 107:703–711
101. Silzel JW, Cercek B, Dodson C et al (1998) Mass-sensing, multianalyte microarray immunoassay with imaging detection. Clin Chem 44:2036–2043
102. Wu P, Hogrebe P, Grainger DW (2006) DNA and protein microarray printing on silicon nitride waveguide surfaces. Biosens Bioelectron 21:1252–1263
103. Bernard A, Michel B, Delamarche E (2001) Micromosaic immunoassays. Anal Chem 73:8–12
104. Delamarche E, Juncker D, Schmid H (2005) Microfluidics for processing surfaces and miniaturizing biological assays. Adv Mater 17:2911–2933
105. Feldstein MJ, Golden JP, Rowe CA et al (1999) Array biosensor: optical and fluidics systems. J Biomed Microdevices 1(2):139–153
106. Golden JP, Shriver-Lake LC, Sapsford KE et al (2005) A "do-it-yourself" array biosensor. Methods 37:65–72
107. Plowman TE, Durstchi JD, Wang HK et al (1999) Multiple-analyte fluoroimmunoassay using an integrated optical waveguide sensor. Anal Chem 71:4344–4352
108. Ziegler J, Zimmermann M, Hunziker P et al (2008) High-performance immunoassays based in through-stencil patterned antibodies and capillary systems. Anal Chem 80:1763–1769
109. Pawlak M, Schick E, Bopp MA et al (2002) Zeptosens' protein microarrays: a novel high performance microarray platform for low abundance protein analysis. Proteomics 2:383–393
110. Zeller PN, Voirin G, Kunz RE (2000) Single-pad scheme for integrated optical fluorescence sensing. Biosens Bioelectron 15:591–595
111. Duveneck GL, Neuschafer D, Ehrat M (1995) Process for detecting evanescently excited luminescence. International Patent Go1N 21/77, 21/64
112. Golden JP, Taitt CR, Shriver-Lake LC et al (2005) A portable automated multianalyte biosensor. Talanta 65:1078–1085
113. Taitt CR, Golden JP, Shubin YS et al (2004) A portable array biosensor for detecting multianalytes in complex samples. Microb Ecol 47:175–185
114. Taitt CR, Anderson GP, Lingerfelt BM et al (2002) Nine-analyte detection using an array-based biosensor. Anal Chem 74:6114–6120
115. Rowe CA, Scruggs SB, Feldstein MJ et al (1999) An array immunosensor for simultaneous detection of clinical analytes. Anal Chem 71:433–439
116. Rowe-Taitt CA, Hazzard JW, Hoffman KE et al (2000) Simultaneous detection of six biohazardous agents using a planar waveguide array biosensor. Biosens Bioelectron 15:579–589
117. Rowe CA, Tender LM, Feldstein MJ et al (1999) Array biosensor for simultaneous identification of bacterial, viral, and protein analytes. Anal Chem 71:3846–3852
118. Ngundi MM, Shriver-Lake LC, Moore MH et al (2006) Multiplexed detection of mycotoxins in foods with a regenerable array. J Food Prot 69:3047–3051
119. Sapsford KE, Ngundi MM, Moore MH et al (2006) Rapid detection of foodborne contaminants using an array biosensor. Sensors Actuators B Chem 113:599–607
120. Moreno-Bondi MC, Taitt CR, Shriver-Lake LC et al (2006) Multiplexed measurement of serum antibodies using an array biosensor. Biosens Bioelectron 21:1880–1886

High-Refractive-Index Waveguide Platforms for Chemical and Biosensing

Katrin Schmitt and Christian Hoffmann

Abstract The field of chemical and biosensors based on waveguide technology is rapidly growing, with new developments focusing on higher sensitivity and stability. This key demand is prompting researchers and developers to explore new materials for waveguide sensor systems, with especially high-refractive-index materials as promising components. This chapter gives an overview of different sensor platforms implementing high-refractive-index waveguide materials, with applications in both research and commercial sensor systems. This is accompanied by a theoretical background of waveguide-sensing principles, especially focusing on the key steps to high sensor sensitivities.

Keywords Evanescent field · Label-free · Fluorescence · Biosensor

Contents

K. Schmitt
Fraunhofer Institute for Physical Measurement Techniques, Heidenhofstr. 8, 79110 Freiburg, Germany

C. Hoffmann (✉)
Institute for Bioprocessing and Analytical Measurement Techniques, Rosenhof, 37308 Heilbad Heiligenstadt, Germany
e-mail: christian.hoffmann@iba-heiligenstadt.de

M. Zourob and A. Lakhtakia (eds.), *Optical Guided-wave Chemical and Biosensors I*, 21
Springer Series on Chemical Sensors and Biosensors 7,
DOI 10.1007/978-3-540-88242-8_2,

Abbreviations

CCD	Charge coupled device
DNA	Deoxyribonucleic acid
MZI	Mach-Zehnder-Interferometer
RNA	Ribonucleic acid

Symbols

c	Speed of light in vacuum
d	Thickness, diameter, distance
d_{eff}	Effective thickness of the waveguide
$\vec{D}$	Electric displacement
$\vec{E}$	Electric field
$\vec{H}$	Magnetic field
i	Imaginary unit
k	Wave vector
L	Interaction length
m	Mode number
n	Refractive index
n_{a}	Refractive index of the ambient medium
n_{ad}	Refractive index of the surface adlayer
n_{eff}	Effective refractive index of the waveguide
n_{w}, n_{c}, n_{s}	Refractive index of waveguide, cover and substrate
P	Light intensity
P_{in}	Input power
P_{out}	Output power
t_{ad}	Thickness of the surface adlayer
TE	Transverse electric
TM	Transverse magnetic
Δz	Penetration depth
α	Coupling angle
β_{m}	Propagation constant of the mode m
$\Delta\Gamma$	Mass coverage
ε	Permittivity

$\overline{\varphi}_{cr}$	Critical angle
$\Delta\varphi$	Phase shift
κ	Diffraction order
λ	Wavelength
λ_0	Wavelength in vacuum
Λ	Grating period
μ	Permeability
τ_r	Phase shift upon reflection
ω	Angular frequency

1 Introduction

Nowadays, chemical sensors and biosensors play a crucial part in everyday life. In the wide field of sensors, those based on waveguide-sensing principles are used, for example, for routine analyses, and more generally, for many applications in physics, chemistry, biology, or medical and pharmaceutical tasks [1–3]. The versatility of waveguide-based sensors combined with their high sensitivity has paved the way for such sensor platforms to play a role in the wide range of possible applications – with many more waiting to be explored. Yet, one crucial prerequisite for the success of a technology is sensitivity, and the sensitivity of waveguides is directly linked to their refractive index. Due to this, waveguide materials with a high refractive-index have attracted recent attention towards the development of new sensors. In addition to the refractive index of the waveguide material, the so-called evanescent field can be optimized to reach high sensor sensitivity. The evanescent field is the exponentially decaying part of the guided light wave lying outside the waveguide. Evanescent field sensors, as presented in this chapter, use this effect to measure the interaction between the sample and the evanescent field. Such sensors can operate either in a label-free manner, i.e., they detect the analyte directly without any reporter molecule, or the evanescent field is used to excite fluorophores attached to the analyte, close to the waveguide surface. The advantage of the selective excitation of bound fluorophores within the evanescent field is that the background signal caused by unbound fluorophores is reduced to a large extent.

Here, we focus on different sensor platforms implementing high-refractive-index waveguide materials with applications in research, as well as commercial sensor systems. The chapter starts with a theoretical background of waveguide-sensing principles, describing the important steps to achieve high sensor sensitivities. This is followed by the introduction of common waveguide fabrication technologies, coupling methods, and high-refractive-index waveguide materials. The second part of the chapter is devoted to sensor principles based on high-refractive-index waveguide materials and presents commercially available sensor systems implementing such waveguides.

2 Waveguide Theory

The term "waveguide" commonly refers to a structure, i.e., a dielectric medium that is able to confine and guide electromagnetic waves. In the simplest case, the waveguide is a material having a refractive index sufficiently high compared to the ambient medium, and appropriate dimensions, to guide light at distinctive wavelengths by total internal reflections. This principle applies to both planar waveguides and optical fibers, which are known from telecommunications applications. Figure 1 shows the basic setup of planar and fiber optic waveguides. Sensor systems for chemical and biosensing applications implement both types of waveguides, depending on the setup and application.

Based on their dimensions and light-guiding properties, both planar and fiber optical waveguides can be subdivided into two classes, namely, *single-mode* (small waveguide thickness) and *multimode* (comparably large thickness). A light wave reproducing itself after two reflections in the waveguide is called an eigenmode, or simply, mode of a waveguide [4]. In single-mode waveguides, only one light mode can be guided, whereas thicker waveguides allow several modes. For multimode waveguides, a description based on the ray-optics approach is adequate. Yet, it does not suffice to describe thin-film (thin core) waveguides, where the electromagnetic

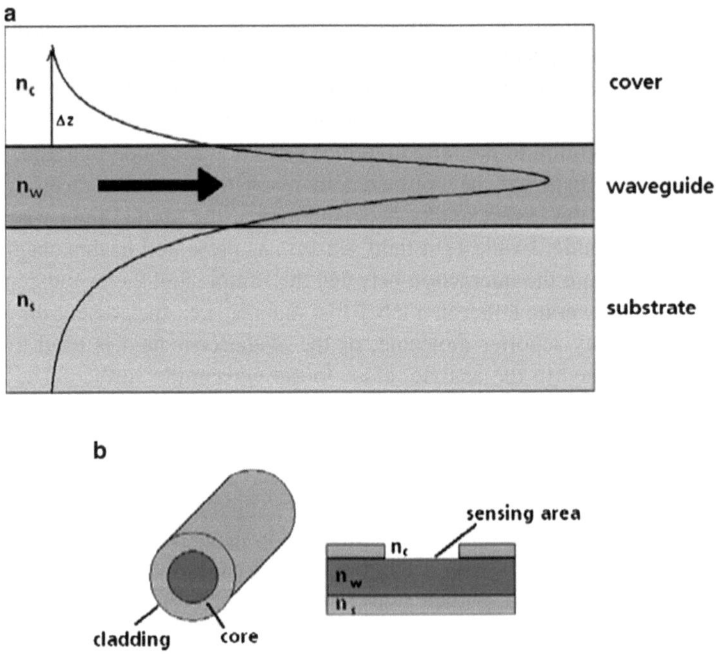

Fig. 1 (**a**) Planar waveguide with refractive index n_w on top of a substrate (refractive index n_s). The evanescent field (penetration depth Δz) of a guided light mode extends into the cover with refractive index n_c. (**b**) Optical fiber waveguide. The same conditions apply as to the planar waveguide

approach is more suitable. These two approaches, which are presented in this chapter, basically follow the description of the optical waveguide theory in Snyder and Love [5]. We exclude a description of laterally structured waveguides, which can be found, for example, in the publications by Kogelnik [6].

2.1 Ray-Optics Approach

Light propagation in optical waveguides is exactly described by Maxwell's equations. Yet, for multimode waveguides, which obey the condition $\frac{2\pi d}{\lambda_0}\sqrt{n_{\mathrm{w}}^2 - n_{\mathrm{c/s}}^2} >> 1$ (d = thickness of the waveguide, diameter for fiber waveguides, λ_0 = wavelength of guided light in vacuum, $n_{\mathrm{w,c/s}}$ = refractive index of waveguide, cover, and substrate, respectively; cf. Fig. 1), the classical geometric approach using ray optics provides a good approximation, and can be based either on direct ray tracing along the waveguide, neglecting any wave effects or on a reduction of the solutions of Maxwell's equations to geometric optics. Here, we focus on the ray tracing method and refer to Snyder and Love [5] for further reading. For simplicity, we assume in the following that the waveguide is planar, with no losses due to absorbance or light scattering (*ideal* waveguide) and an unbounded substrate/cover. In an ideal waveguide, light could therefore propagate over arbitrarily large distances with no loss of power. Furthermore, only step-index waveguides with n_{w} = const. are considered, as shown in Fig. 2.

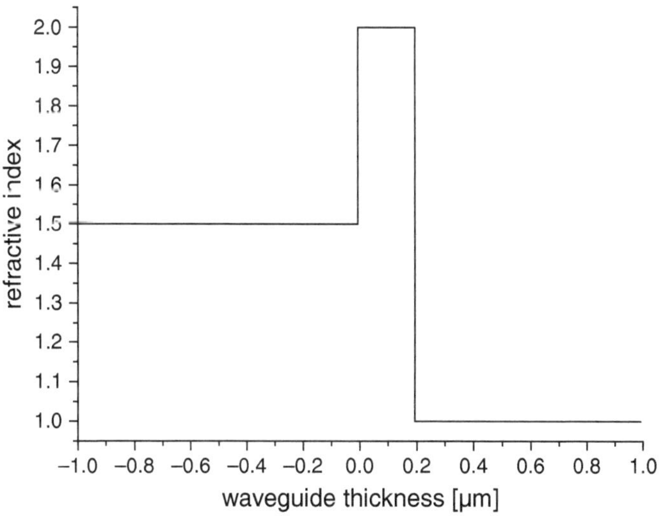

Fig. 2 Refractive index profile of a 200 nm step-index waveguide with refractive indices of $n_{\mathrm{s}} = 1.5$, $n_{\mathrm{w}} = 2.0$ and $n_{\mathrm{c}} = 1.0$

We now consider a light wave with $\lambda = \lambda_0/n$, polarized in the x-direction, traveling in the waveguide between the two interfaces, waveguide-cover and waveguide-substrate. The wave will be reflected at the interfaces with angles φ_c and φ_s for cover and substrate, respectively. Since we assume the electric field is parallel to the interfaces, a phase shift of τ_r will occur at each interface according to Fresnel's formulas, while the amplitude and polarization remain the same. Additionally, we assume for simplicity that $\varphi_c = \varphi_s = \varphi$.

We now consider only waves that reproduce themselves after two reflections (resonance condition), called *eigenmodes* (or simply *modes*) of the waveguide. These modes have the same transversal field distribution and polarization along the waveguide, which is a direct result of the resonance condition [4]. Figure 3 represents the resonance condition graphically. If we now consider the resonance condition with the help of Fig. 3, i.e., the phase difference needs to be a multiple of 2π, we get

$$\frac{2\pi}{\lambda}\overline{AB} = \frac{2\pi}{\lambda}\overline{AC} - 2\pi m - 2\tau_r \quad \text{with } m = 1, 2, 3, \ldots \text{ (mode number)} \tag{1}$$

After two reflections, the reflected wave is at a distance of $\overline{AC} - \overline{AB} = 2d \sin\varphi$ from the original wave:

$$\frac{2\pi}{\lambda} 2d \sin\varphi - 2\tau_r = 2\pi m \quad \text{with} \quad m = 1, 2, 3, \tag{2}$$

The phase shift τ_r upon reflection at the interfaces depends on both the angle of reflection φ and on the polarization of the electromagnetic wave (TE or TM).

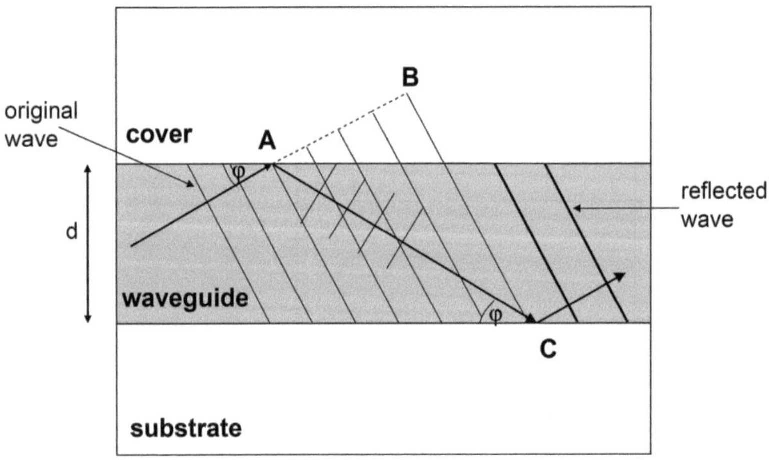

Fig. 3 Geometrical representation of the resonance condition. The phase difference after two reflections at the boundary layers has to be a multiple of 2π. Additionally, a phase shift of τ_r occurs at each interface

For TE (electric field is perpendicular to the plane of incidence spanned by the wave normal and the normal to the interface), the phase shift is

$$\tan\frac{\tau_r}{2} = \frac{\sqrt{\sin^2\overline{\varphi} - \sin^2\overline{\varphi}_{cr}}}{\cos\overline{\varphi}}, \tag{3}$$

where $\overline{\varphi}_{cr}$ is the critical angle for total internal reflection and $\overline{\varphi}$ denotes the angle to the perpendicular. The definition of the *effective refractive index*,

$$n_{eff} = n_w \sin\overline{\varphi}, \tag{4}$$

which is a measure for the phase velocity of the guided light wave, and together with $\sin\overline{\varphi}_{cr} = \frac{n_{c/s}}{n_w}$ and $\cos\overline{\varphi} = \sqrt{n_w^2 - n_{eff}^2}/n_w$ allows to describe the phase shift in the following form:

$$\tau_r = 2\ \mathrm{arc}\tan\left\{\frac{n_w\sqrt{\frac{n_{eff}^2}{n_w^2} - \frac{n_{c/s}^2}{n_w^2}}}{\sqrt{n_w^2 - n_{eff}^2}}\right\} = 2\ \mathrm{arc}\tan\sqrt{\frac{n_{eff}^2 - n_{c/s}^2}{n_w^2 - n_{eff}^2}}. \tag{5}$$

With the wave vector k and its components, $k_x = 0$, $k_y = n_w k \sin\overline{\varphi}$ and $k_z = n_w k \cos\overline{\varphi}$, we can write this expression, using equation (4), as

$$kd\sqrt{n_w^2 - n_{eff}^2} - 2\ \mathrm{arc}\tan\sqrt{\frac{n_{eff}^2 - n_{c/s}^2}{n_w^2 - n_{eff}^2}} = \pi m. \tag{6}$$

After defining

$$a = k\sqrt{n_w^2 - n_{eff}^2}, \tag{7}$$

$$b = k\sqrt{n_{eff}^2 - n_{c/s}^2}, \tag{8}$$

we can rewrite (6) as a transcendental equation for the determination of n_{eff}:

$$\tan(ad - \pi m) = \frac{2ab}{a^2 - 2b}. \tag{9}$$

This expression can be solved either graphically or numerically. The solutions are discrete, i.e., the wave can be guided only for certain values of n_{eff}, corresponding to one mode. A more detailed description of the graphical solution (mode diagram) follows in the next section.

2.2 Electromagnetic Approach

The general description of electromagnetic waves is based on Maxwell's equations. For applications in optics, as that is important here, they reduce to the special case of charge- and current-free media. A subset of the solutions of the Maxwell equations can be assumed to be a harmonic wave of the form:

$$A(t) = A\ \mathrm{e}^{i\omega t}, \tag{10}$$

with the angular frequency $\omega = 2\pi c/\lambda$. Inserting this representation and $\vec{D} = \varepsilon\vec{E}$ into Maxwell's equations gives

$$\vec{\nabla} \times \vec{E} = -i\omega\mu\vec{H}, \tag{11}$$

$$\vec{\nabla} \times \vec{H} = i\omega\varepsilon\vec{E}, \tag{12}$$

with

μ = permeability
$\vec{E}$ = electric field
$\vec{H}$ = magnetic field
$\vec{D}$ = electric displacement
ε = permittivity

In the case of a planar waveguide with z as the direction of light propagation, the solutions to these equations can be described in the following form:

$$A(x, y, z) = A_{\mathrm{m}}(x, y)e^{-i\beta_{\mathrm{m}} z}, \tag{13}$$

with $\beta_{\mathrm{m}} = \frac{2\pi}{\lambda_0} n_{\mathrm{w}} \sin\varphi$ being the propagation constant and m the mode index. The mode index is omitted now for simplicity; the different modes of a waveguide are discussed later in greater detail. Inserting (13) in (12) and (11) and separating x, y, and z yields

$$\frac{\partial E_z}{\partial y} + i\beta E_{\mathrm{y}} = -i\omega\mu H_{\mathrm{x}}, \tag{14}$$

$$i\beta E_{\mathrm{x}} + \frac{\partial E_z}{\partial x} = i\omega\mu H_{\mathrm{y}}, \tag{15}$$

$$\frac{\partial E_y}{\partial x} - \frac{\partial E_x}{\partial y} = -i\omega\mu H_z, \tag{16}$$

$$\frac{\partial H_{\mathrm{z}}}{\partial y} + i\beta H_{\mathrm{y}} = i\omega\varepsilon E_{\mathrm{x}}, \tag{17}$$

$$i\beta H_x + \frac{\partial H_z}{\partial x} = -i\omega\varepsilon E_y, \quad (18)$$

$$\frac{\partial H_y}{\partial x} - \frac{\partial H_x}{\partial y} = i\omega\varepsilon E_z. \quad (19)$$

As stated earlier, the direction of light propagation is z, and the confinement of the waveguide is in the x-direction. It is assumed that the waveguide is not structured laterally, i.e., has an infinite extension in the y-direction. Then, without loss of generality, we can assume that $\frac{\partial}{\partial y} \equiv 0$, because under this condition, the electric and magnetic fields supported by the waveguide do not depend on the y-direction.

From these equations, the mode distribution in waveguides can be calculated. As stated earlier, light waves satisfying the resonance condition, i.e., reproducing themselves after two reflections in the waveguide, are called modes. The mode equation for three-layer planar waveguides has been described by Tiefenthaler and colleagues in the following form [7]:

$$2\,kd\sqrt{n_w^2 - n_{\mathrm{eff}}^2} + \varphi_c + \varphi_s = 2\pi m. \quad (20)$$

It can be seen directly from the mode equation where $k = \frac{2\pi}{\lambda_0}$ that with an increasing ratio of thickness d to the wavelength λ, more modes can be guided. In a planar waveguide, where linearly polarized light is guided, two different modes exist: the TE mode supports only electrical fields perpendicular to the direction of propagation ($E_z = 0$), similarly to the TM mode, where $H_z = 0$. Considering this for equations (14)–(19), we get for TE modes:

$$\beta E_y = -\omega\mu H_x, \quad (21)$$

$$\frac{\partial E_y}{\partial x} = -i\omega\mu H_z, \quad (22)$$

$$\frac{\partial H_z}{\partial x} + i\beta H_x = -i\omega\varepsilon E_y, \quad (23)$$

and for TM:

$$i\beta E_x + \frac{\partial E_z}{\partial x} = i\omega\mu H_y, \quad (24)$$

$$\beta H_y = \omega\varepsilon E_x, \quad (25)$$

$$\frac{\partial H_y}{\partial x} = i\omega\varepsilon E_z, \quad (26)$$

where β is the propagation constant. Guided modes, i.e., waves undergoing total internal reflection, are bounded by the following condition:

$$kn_{c/s} \le \beta \le kn_w. \tag{27}$$

Inserting this boundary condition into (21)–(26) yields for TE:

$$E_y = Ae^{-\gamma x} \quad \text{for} \quad x \ge 0, \tag{28}$$

$$E_y = A\left(\cos(\delta x) - \frac{\gamma}{\delta}\sin(\delta x)\right) \quad \text{for} \quad 0 \ge x \ge -d, \tag{29}$$

$$E_y = A\left(\cos(\delta h) + \frac{\gamma}{\delta}\sin(\delta h)\right)e^{\alpha(x+h)} \quad \text{for} \quad -d \ge x, \tag{30}$$

and for TM:

$$H_y = Ae^{-\gamma x} \quad \text{for} \quad x \ge 0, \tag{31}$$

$$H_y = A\left(\cos(\delta x) - \left(\frac{n_w}{n_c}\right)^2 \frac{\gamma}{\delta}\sin(\delta x)\right) \quad \text{for} \quad 0 \ge x \ge -d, \tag{32}$$

$$H_y = A\left(\cos(\delta h) + \left(\frac{n_w}{n_c}\right)^2 \frac{\gamma}{\delta}\sin(\delta h)\right)e^{\alpha(x+h)} \quad \text{for} \quad -d \ge x, \tag{33}$$

with A = arbitrary constant, and

$$\alpha = \sqrt{\beta^2 - k^2 n_s^2}, \tag{34}$$

$$\gamma = \sqrt{\beta^2 - k^2 n_c^2}, \tag{35}$$

$$\delta = \sqrt{k^2 n_w^2 - \beta^2}. \tag{36}$$

To illustrate these results, Fig. 4 shows an example for a normalized field distribution H_y for a TM mode with index 0.

At the interfaces between the media, the tangential component of E_y as well as the tangential component of H_y needs to be continuous. These transition conditions result in a set of linear equations which only yield nontrivial solutions if the coefficient determinant vanishes, and we get the eigenvalue:

$$\tan(\delta d - m\pi) = \frac{\delta(r_c\gamma + r_s\gamma)}{\delta^2 - r_c r_s \gamma\alpha}, \tag{37}$$

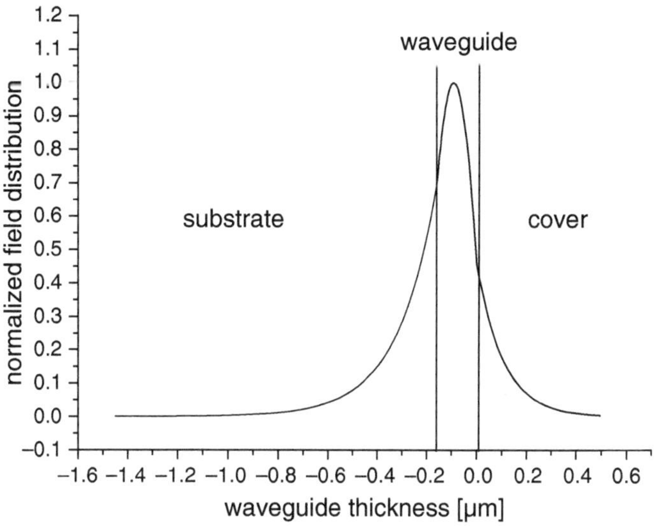

Fig. 4 TM-mode field distribution for a planar waveguide with refractive indices of $n_s = 1.52$, $n_w = 2.1$, $n_c = 1.333$ at $\lambda_0 = 675$ nm, and $m = 0$ [8]

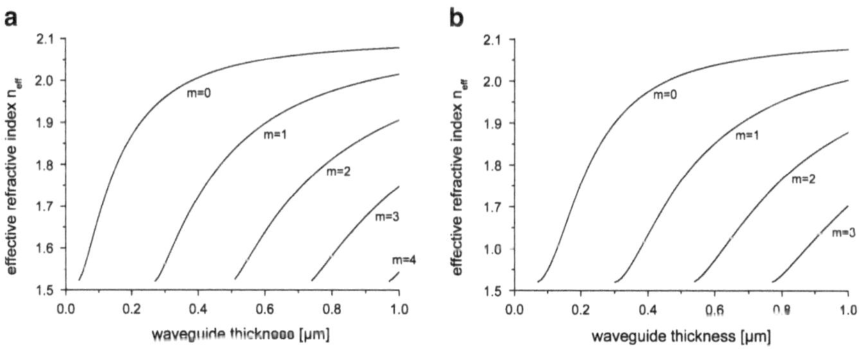

Fig. 5 Mode diagrams for a dielectric planar waveguide with refractive indices of $n_s = 1.52$, $n_w = 2.1$, $n_c = 1.333$ at $\lambda_0 = 675$ nm. (**a**) TE modes, (**b**) TM modes [8]

with $r_{c/s} = 1$ for TE modes and $r_{c/s} = \left(\frac{n_s}{n_c}\right)^2$ for TM modes. This eigenvalue equation can be solved either graphically or numerically, and has already been found with the ray-optics approach. Figure 5 shows mode diagrams for TE and TM modes using the following parameters: $n_s = 1.52$, $n_w = 2.1$, $n_c = 1.333$ at $\lambda_0 = 675$ nm [8].

When the waveguides are used as sensors, i.e., in label-free applications, their sensitivity to changes in the cover layer is important. In the following, again ideal waveguides will be treated, i.e., nonporous ones, so that they react only to changes on their surface, and diffusion processes into the waveguide layer will be neglected. Two types of surface changes and, thus, the effective refractive index can occur: either, the bulk refractive index changes, e.g., variations in the refractive index of the buffer, or molecules adsorb or bind to the surface, also resulting in a signal change of the effective refractive index. The general expression for changes in n_{eff} is given by Tiefenthaler and Lukosz [7]:

$$\Delta n_{\text{eff}} = \left(\frac{\partial n_{\text{eff}}}{\partial t_{\text{ad}}}\right)\Delta t_{\text{ad}} + \left(\frac{\partial n_{\text{eff}}}{\partial n_{\text{c}}}\right)\Delta n_{\text{c}} + \left(\frac{\partial n_{\text{eff}}}{\partial n_{\text{w}}}\right)\Delta n_{\text{w}}, \tag{38}$$

where the last summand can be omitted for nonporous waveguides. t_{ad} denotes the thickness of the surface adlayer. It is assumed that the molecules form a homogeneous layer with $t_{\text{ad}} << \lambda_0$.

From the mode equation and equation (38), the waveguide sensitivity for both cover refractive index and the surface adlayer changes can be derived (three-layer planar waveguide). The derivation is given in detail in [7] and here only the results are given:

$$\frac{\partial n_{\text{eff}}}{\partial n_{\text{c}}} = \left(\frac{n_{\text{c}}}{n_{\text{eff}}}\right)\left(\frac{n_{\text{w}}^2 - n_{\text{eff}}^2}{n_{\text{w}}^2 - n_{\text{c}}^2}\right)\left(\frac{\Delta z_{\text{c/s}}}{d_{\text{eff}}}\right)\left[2\left(\frac{n_{\text{eff}}}{n_{\text{c}}}\right)^2 - 1\right]^{\rho} \tag{39}$$

for cover refractive index changes, and

$$\frac{\partial n_{\text{eff}}}{\partial t_{\text{ad}}} = \left(\frac{n_{\text{w}}^2 - n_{\text{eff}}^2}{n_{\text{eff}}^2\, d_{\text{eff}}}\right)\left(\frac{n_{\text{ad}}^2 - n_{\text{c}}^2}{n_{\text{w}}^2 - n_{\text{c}}^2}\right)\left[\frac{\left(\frac{n_{\text{eff}}}{n_{\text{c}}}\right)^2 + \left(\frac{n_{\text{eff}}}{n_{\text{ad}}}\right)^2 - 1}{\left(\frac{n_{\text{eff}}}{n_{\text{c}}}\right)^2 + \left(\frac{n_{\text{eff}}}{n_{\text{w}}}\right)^2 - 1}\right]^{\rho} \tag{40}$$

for surface adlayer changes, where $\Delta z_{\text{c/s}}$ are the penetration depths of the evanescent field into cover (c) and substrate (s), respectively, n_{ad} denotes the refractive index of the surface adlayer, $\rho = 0$ for TE modes and $\rho = 1$ for TM modes, and

$$d_{\text{eff}} = d + \Delta z_{\text{c}} + \Delta z_{\text{s}} \tag{41}$$

defines the effective thickness of waveguide.

Equations (39) and (40) can again be solved by numerical iteration. Figure 6 shows the sensitivities $\frac{\partial n_{\text{eff}}}{\partial n_{\text{c}}}$ and $\frac{\partial n_{\text{eff}}}{\partial t_{\text{ad}}}$ for the first two TE and TM modes (0 and 1). The calculations show higher sensitivities for TM modes, while a waveguide thickness of 150–160 nm yields the optimum sensitivity. The equations also imply that high sensitivities can be reached with monomode waveguides, having a high refractive index difference to the substrate refractive index.

In other cases, when the light in the evanescent field of the waveguides is used to excite fluorophores on the waveguide surface, the light intensity in the evanescent

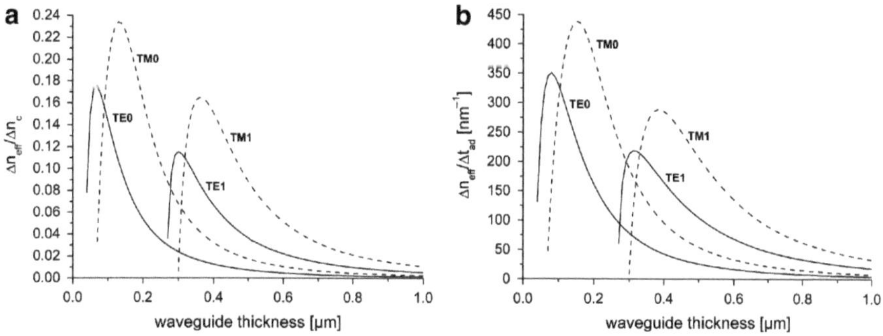

Fig. 6 Theoretical sensitivities of waveguides depending on the waveguide thickness, (**a**) to cover refractive index changes, (**b**) to surface adlayer changes with $n_s = 1.52$, $n_w = 2.1$, $n_c = 1.333$ at $\lambda_0 = 675$ nm [8]

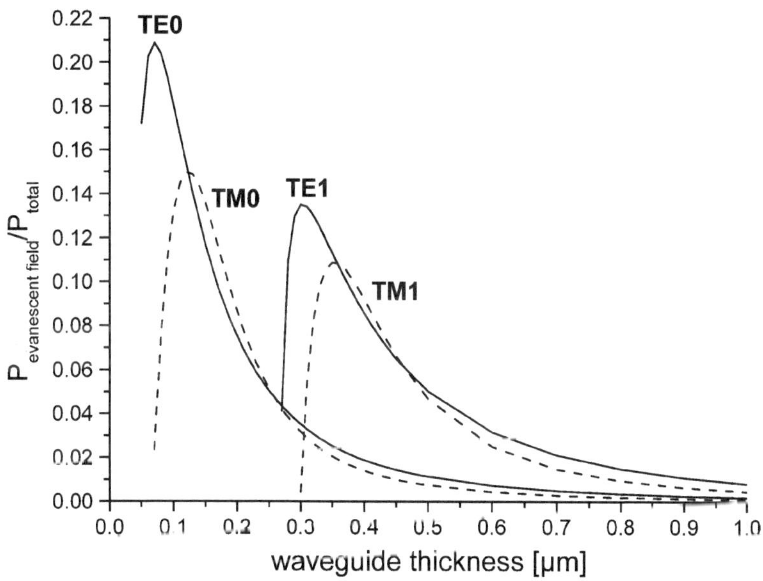

Fig. 7 Theoretical light intensities in the evanescent field as the fraction of the total light intensity in the waveguide versus waveguide thickness. Parameters for calculation: $n_s = 1.52$, $n_w = 2.1$, $n_c = 1.333$ at $\lambda_0 = 675$ nm

field is crucial to optimize the excitation efficiency. Figure 7 depicts the light intensity in the evanescent field as a fraction of the total light intensity in the waveguide, depending on the waveguide thickness, according to the following equation:

$$\frac{P_{\text{evanescent field}}}{P_{\text{total}}} = \left(\frac{n_w^2 - n_{\text{eff}}^2}{n_w^2 - n_c^2}\right)\frac{\Delta z_c}{d_{\text{eff}}}. \tag{42}$$

The TE0 mode provides the highest light intensity in the evanescent field for monomode waveguides with thicknesses from 60 to 70 nm.

3 Waveguide Fabrication Technologies

To achieve high-refractive-index optical waveguides, the appropriate waveguide technology has to be chosen. For chemical and biosensing applications, not only sensitivity is a crucial prerequisite, but also an easy structuring process and the resistance against a wide range of chemicals used for bio-molecule immobilization and their regeneration.

Several technologies have evolved to fabricate optical waveguides, which can be divided into two classes: in the first case, the substrate material is modified, e.g., by ion exchange in SiO_2 or $LiNbO_3$ [9, 10]. These waveguides have the disadvantage that their refractive index is increased only to a minor extent, yet they show less porosity than coated waveguide layers. In the second case, a waveguide layer is applied on top of a substrate material (coating) as it is done with metal oxides or nitrides [11–14]. Here, high-refractive-index waveguides can be achieved more easily, but often they are porous and lack sufficient chemical resistance. These waveguides are mainly fabricated using chemical vapor deposition techniques (CVD). The need for low porosity in combination with high chemical resistance requires optimized coating techniques to fabricate extremely dense waveguide films, also to reduce surface roughness to minimize light losses due to scattering.

4 Light Coupling

Because of the small dimensions of thin-film waveguides, the coupling of light requires special attention. The application of the sensor system and the layout of the waveguide determine the light coupling method. Figure 8 shows four commonly used coupling methods for thin-film waveguides.

Front-face coupling can be accomplished either by a lens focusing a collimated light beam onto the waveguide or by a waveguide illumination with an optical fiber. Effective coupling requires well-prepared (e.g., polished) square edges of the waveguide and an exact alignment of the optical elements focusing the light beam. This ensures a maximal overlap integral, which is determined by the intensity distribution of the light and the distribution of the waveguide modes. Although technically not demanding, front-face coupling has the disadvantage of low robustness against vibrations in the sensor system, and it requires extensive alignment procedures to minimize variations in the coupling efficiency.

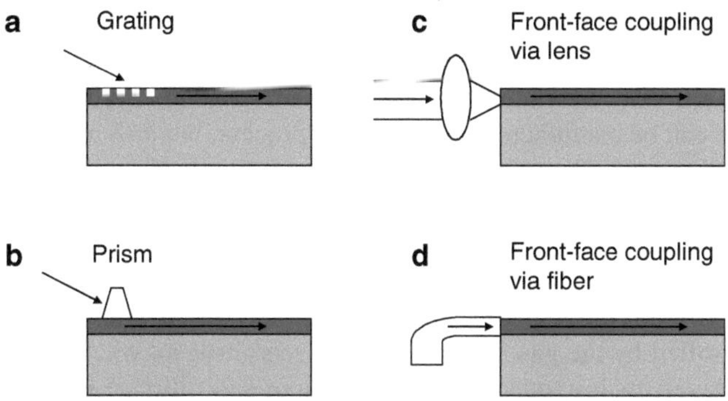

Fig. 8 Light coupling methods into waveguides. (**a**) Grating coupling, (**b**) prism coupling, (**c**) front-face coupling via optical lens (**d**) front-face coupling via optical fiber

Prism coupling also requires no waveguide structuring and is a relatively simple method. Light is coupled into the waveguide via an attached prism on the waveguide surface. Compared to front-face coupling, the alignment of the light beam is not as critical as long as the correct coupling angle is used. The coupling angle depends on the effective refractive index of the waveguide.

However, this coupling method might not be suitable for sensor systems used outside the laboratory, since the pressure of the prism on the waveguide needs to be reproduced with high precision.

For grating coupling, a defined grating structure in the waveguide is necessary, which is technically more elaborate than the other methods. As with prism coupling, the coupling angle is determined by the effective refractive index of the waveguide (sensing principle of the grating coupler sensor). The grating structure allows only a very narrow range of coupling angles, depending on the spectral bandwidth of the light source. The position of the light beam on the grating influences the coupling efficiency to a large extent, along with grating parameters, i.e., structure depth. When linearly polarized light is used, the direction of polarization, i.e., TE or TM mode, is decisive. An advantage of this coupling method over the previously mentioned methods is a good reproducibility of coupling conditions since no further optical elements are directly involved.

5 Waveguide Materials

Materials that enable light mode propagation are characterized by a high refractive index and a low attenuation <3 dB/cm. Especially, metal oxides or nitrides meet these optical requirements, and hence SiO_2, SiO_xN_y, Si_3N_4, TiO_2, Ta_2O_5, and

Nb_2O_5 are the typical waveguide materials. The first waveguides used for sensor applications were based on silicon oxide–titanium oxide (SiO_2–TiO_2) ($n_{632.8} = 1.750$–2.0 depending on the titanium content, normally $n_{632.8} = 1.8$ [15]). These can be manufactured by a sol-gel process, but also magnetron-sputtering was applied [16]. The refractive index of pure titanium oxide TiO_2 is $n \approx 2.7$.

Based on thin-film technology, integrated optical waveguiding chips can be produced with silicon oxynitride SiO_xN_y. Depending on the oxygen and nitrogen content, the refractive index can be varied between pure silicon oxide SiO_2 ($n_{632.8} = 1.46$) and pure silicon nitride Si_3N_4 ($n_{632.8} = 2.01$). The specific value can be controlled by the gas flow ratio of oxygen/nitrogen, when depositing these waveguide layers by low-pressure chemical vapor deposition (LPCVD) or plasma-enhanced chemical vapor deposition (PECVD) technologies [17].

Even higher refractive indices for waveguide materials are given for tantalum pentoxide Ta_2O_5 ($n_{632.8} = 2.22$ [18] or between $n = 2.1$ and $n = 2.2$ for the red spectral range [19, 20]). These films are commonly deposited by a reactive sputtering process. Niobium oxide Nb_2O_5 ($n = 2.36$) is also used as planar-waveguide material [21, 22].

The refractive indices cited for the waveguide materials are approximate values and are usually determined for the He–Ne laser wavelength 632.8 nm or for the red spectral range. Depending on the purity and the porosity of the waveguide material, the refractive index may vary.

6 Sensor Principles Based on High-Refractive-Index Optical Waveguides

High-refractive-index materials enable the investigation of bioreactions at the surface within the evanescent field. Various sensor systems were developed with the advantage of high sensitivity and measurement in the presence of the sample without any rinsing. The sensor principles can be divided into label-free sensor systems where the determination of the effective refractive index is the pivotal parameter and systems based on surface-confined fluorescence excitation where the marker molecule is solely detected.

Systems based on the monitoring of refractive-index changes are suitable for the determination of affinity constants and kinetic studies. Grating coupler and interferometric devices were developed for this purpose. The advantage of simple sample preparation without reporter molecule faces the disadvantage of interfering signals caused by nonspecific binding effects of matrix molecules, since the signal depends mainly on the mass coverage on the surface without any discrimination of molecules. Evanescent field fluorescence systems are characterized by the independence of the molecular mass of the target molecule since they only detect the fluorescent label disregarding the properties of the molecule of interest.

6.1 Grating-Based Label-Free Detection Systems

6.1.1 Grating Coupler

Various gratings with different design have been developed for the energy transfer of a light beam into or out of an optical waveguide. They are characterized by their geometrical dimensions and the grating period (Fig. 9). The different shapes encompass triangular, rectangular, and trapezoidal profiles, but asymmetric profiles are also possible [23]. The measurement principle of the grating coupler for chemical sensing was first discovered and published by Lukosz and Tiefenthaler [24]. They investigated the coupling of a He–Ne laser beam ($\lambda = 632.8$ nm) into an embossed surface relief grating on a planar waveguide consisting of SiO_2 and TiO_2. They observed a propagation of the wave within the waveguide if the coupling condition (43) is fulfilled. Yet, this condition changes in the presence of different ambient gases or humidity, and this effect is reversible. It can, thus, also be exploited in such a way that the ambient conditions are adapted to the coupling condition. This phenomenon is explained by an increasing effective thickness of the waveguide or rather the effective refractive index by water and/or gas adsorption. The coupling always occurs at a certain angle of the incident light beam depending on the effective refractive index of the waveguide. In the identical manner, the coupling condition is valid for outcoupling:

$$n_a \sin \alpha = \kappa \frac{\lambda_0}{\Lambda} - n_{\mathrm{eff}}, \tag{43}$$

with n_a being the refractive index of the ambient medium (air, buffer), α the coupling angle, κ the diffraction order, λ_0 the wavelength in vacuum, and Λ the grating period.

Therefore, various detection principles are applicable to measure the adsorption of a thin layer on a waveguide grating (Fig. 10). This layer may consist of a gas

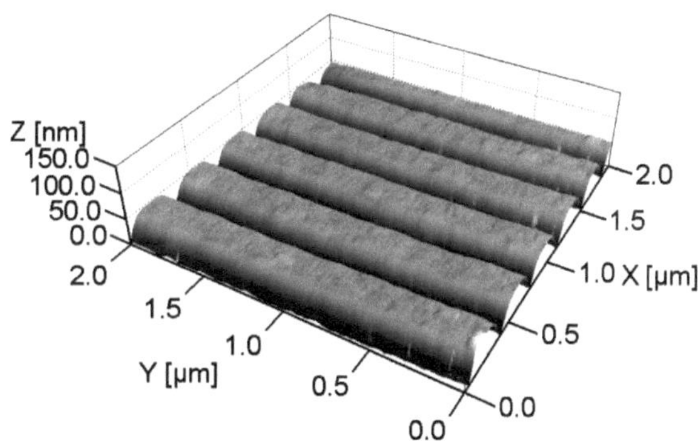

Fig. 9 Image of a grating area of 2 μm × 2 μm recorded by atomic force microscopy. Grating depth ~30 nm, period ~330 nm

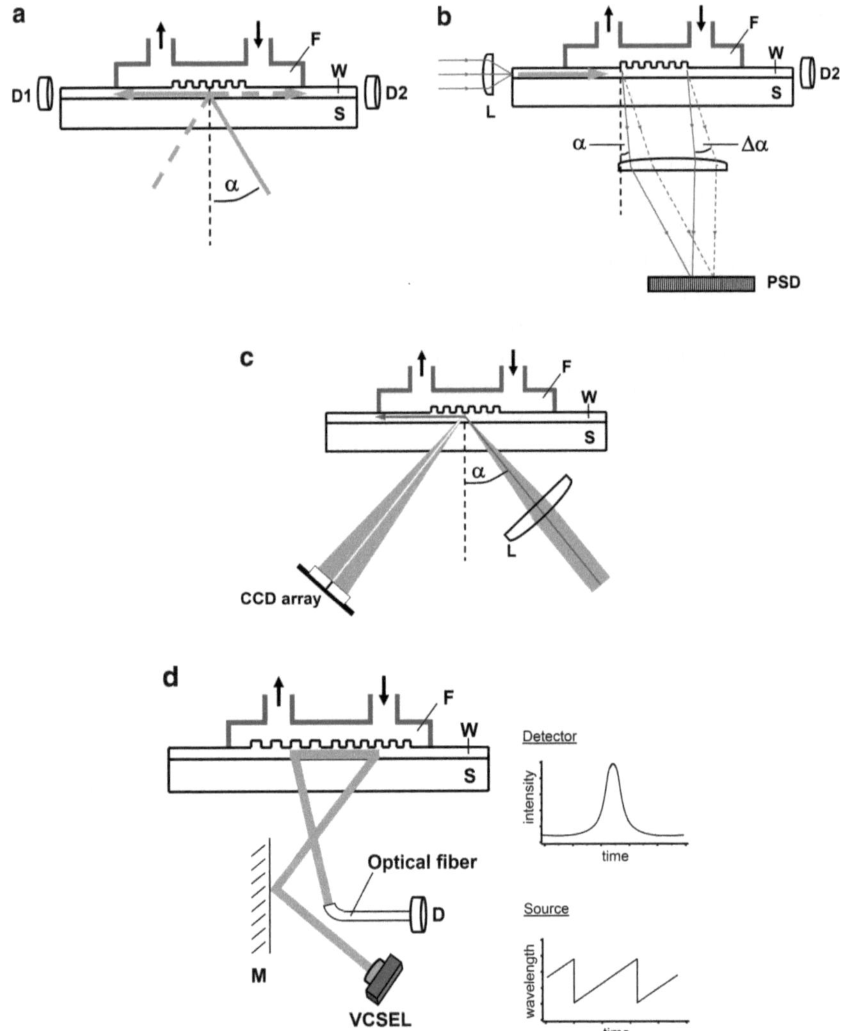

Fig. 10 Schematic drawings of the various principles of grating coupler systems. (**a**) input grating coupler [25], (**b**) output grating coupler [28], (**c**) reflected-mode grating coupler [30], (**d**) wavelength interrogated optical sensor WIOS [19]; *S* substrate, *W* waveguide, *F* fluidic chamber, *L* lens, *PSD* position sensitive detector, *CCD* charge coupled device, *D* photodiode, α coupling angle, $\Delta\alpha$ change in coupling angle, *M* mirror (movable), *VCSEL* vertical cavity surface-emitting laser diode

(at different humidity) or any kind of organic matter including biological molecules. The monitoring of the coupling angle of a laser beam into a waveguide via a grating is called *input grating coupler* [25, 26]. A photodetector at the end of the planar waveguide detects the coupling efficiency. The sample can be rotated on a goniometer to find the coupling angle. The optical arrangement consists only of a coherent light source, collimating lenses, and a photodetector, e.g., a silicon photodiode. The reverse principle, i.e., the detection of the angle of the light beam that propagates first within a waveguide and is then coupled out via a grating, is utilized by the *output grating coupler* [27, 28]. Here, light is coupled into the horizontally positioned waveguide and a position-sensitive photodetector, e.g., a CCD line sensor, monitors the coupling angle depending on the layer thickness or the effective refractive index, respectively. The detected signal then represents the position of the outcoupled light beam on the detector. This arrangement avoids moving parts, leading to a higher signal stability, and therefore higher sensitivity. Another alternative is the *reflection mode grating coupler* that applies an angle spectrum of the light beam to the grating [18, 29, 30]. A collimating lens provides a continuum of angles undergoing reflection except the angle that is obeying the coupling condition. This leads to a dark line corresponding to the coupling angle at the effective refractive index in the reflected light, which is detected by an array detector, e.g., a CCD camera. The latter method suits best for a parallel readout of signals and is, therefore, most promising for high throughput applications and commercialization of such sensor systems since there are no moving parts necessary for the optical readout.

While the aforementioned methods are based on the detection of the coupling angle at a constant wavelength, it is also feasible to keep the incident angle constant and scan the wavelength with a system which is called the *wavelength interrogated optical sensor* (WIOS) [19]. A vertical cavity surface-emitting laser diode (VCSEL) with a saw-tooth wavelength modulation of 2 nm serves as a light source that is directed towards the grating by an adjustable mirror. The coupling condition is fulfilled at a certain wavelength combined with a fixed angle. The light mode is guided through the waveguide and is coupled out by a second grating. Then, the light beam is collected by an optical fiber and detected by a photodiode. The resonance wavelength changes with refractive index variations of the adjacent medium on the surface of the waveguide.

6.1.2 Colorimetric Resonant Grating Reflection

Another approach that is also suitable for a parallel readout of binding events is *colorimetric resonant reflection* at a grating with subwavelength structures (Fig. 11) [31, 32]. This is manufactured from a high-refractive-index material and sandwiched between a substrate with lower refractive index, e.g., silicon oxide (glass) and a cover layer that fills the grating grooves, with a lower refractive index as well. The grating exhibits a very small period so that only a sharp wavelength maximum is reflected upon illumination with white light. Consequently, the optical setup consists only of a white light source that illuminates a small spot of the substrate with the grating structure via an optical fiber at normal incidence and a spectrometer that detects the peak wavelength collected by a second fiber, also at normal incidence. When molecules bind to the grating surface, the reflected light shifts

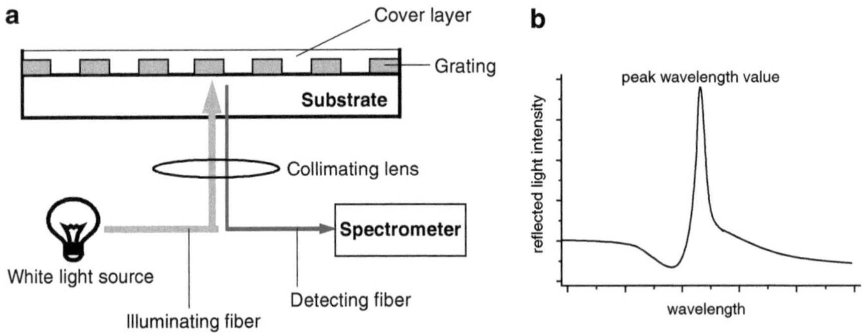

Fig. 11 (**a**) Schematic drawing of the colorimetric resonant reflection sensor. White light illuminates the sensor surface by an optical fiber. A second fiber guides the reflected light to a spectrometer. (**b**) Schematic spectrum

its wavelength due to the change of the optical path of light that is coupled into the grating. Hence, biological molecules like peptides, proteins, or DNA can be detected if the complementary capture molecule is immobilized. Since no coupling optics such as prisms is necessary, the gratings can be implemented in a planar substrate like microtiter plates. This makes it a promising potential high throughput detection system, which is now commercialized by Corning Inc. [21] or SRU Biosystems [33].

6.2 Interferometric Systems

Interferometric methods are based on the superposition of two coherent light beams, where one beam has interacted with the sample and the other serves as reference. The superposition of both beams leads to constructive or destructive interferences. Depending on the modulation of the light source, these interferences can be detected simply by a photodiode or laterally resolved by a CCD detector. A variation of the refractive index within the evanescent field, e.g., owing to adsorption of the sample to the waveguide surface, results in a change of the effective refractive index of the waveguide and hence in a phase shift $\Delta\varphi$, which can be described as follows:

$$\Delta\varphi = \frac{2\pi}{\lambda_0} L \Delta n_{\mathrm{eff}}, \tag{44}$$

where λ_0 denotes the vacuum wavelength and L the interaction length, i.e., the distance the evanescent field is in contact with the sample. Depending on the experimental approach, the reference branch can compensate intrinsic instabilities due to light source or ambient temperature fluctuations. The most investigated approach is the so-called *Mach-Zehnder-Interferometer* (MZI) configuration [17, 34]. It consists of one light beam that is split into a sensing and a reference branch. While the sensing branch interacts with the sample, the reference branch is in contact with the reference medium, e.g., air, buffer, solvent, etc., and eventually both branches are combined to interfere. Applying this MZI configuration to a sensor chip means that this structure has to be deposited with a high-refractive-index material on a substrate. When monochromatic light is guided through this waveguide structure, the ratio of the beam power is exactly one and the phase difference is exactly zero if the sensor chip is perfectly manufactured. Applying a sample to the sensing branch, the local refractive index will change at the waveguide surface inducing a phase shift, and the output power will change obeying the following equation:

$$P_{\text{out}} = \frac{1}{2} P_{\text{in}} (1 + \cos \Delta\varphi). \tag{45}$$

Here, P_{out} denotes the output power, P_{in} the input power, and $\Delta\varphi$ the sample-induced phase shift. The MZI is generally designed as a system using a monomode wave propagating through the waveguide. The phase shift as a measure for the sample concentration can be derived from the power ratio $P_{\text{out}}/P_{\text{in}}$. This implies that in addition to P_{out}, P_{in} also has to be determined by a second photodetector. Since the power ratio is a sinusoidal function of $\Delta\varphi$, there are the following ambiguities (Fig. 12) inherent in the MZI:

- The fringe order ambiguity: the periodical function leads to equally probable solutions of $\Delta\varphi$ for a measured value of $P_{\text{out}}/P_{\text{in}}$: $\Delta\varphi' = \Delta\varphi + k2\pi$, with k being an integer.
- The directional ambiguity: starting on the sinusoidal function from one extremum, the sign of the slope for a $\Delta\varphi$ value cannot be deduced leading to different $\Delta\varphi$ values for one $P_{\text{out}}/P_{\text{in}}$ value depending on the location on the right- or left-hand side of the extremum.

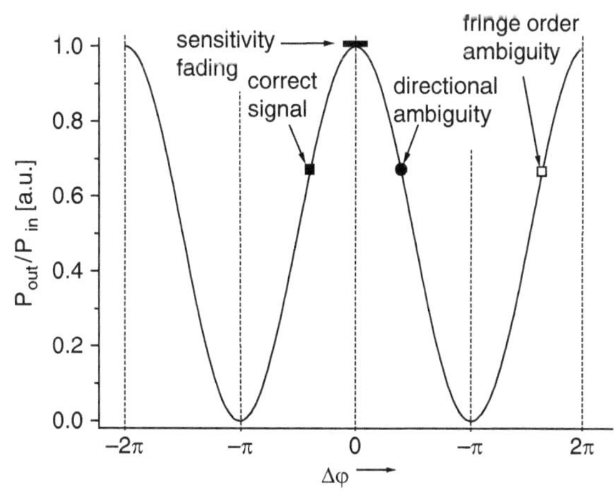

Fig. 12 Output/input power ratio $P_{\text{out}}/P_{\text{in}}$ as a function of the phase difference $\Delta\varphi$ between the sensing and the reference branch illustrating the inherent problems of determining $\Delta\varphi_{\text{m}}$

- Sensitivity fading: since the slope at the extrema is zero, for P_{out}/P_{in} values equal to zero or one, $\Delta\varphi$ values are not determinable with high sensitivity. The highest sensitivity can be obtained at the points with the steepest slope at $\Delta\varphi = (n + ½)\pi$ with n being an integer.

The problems of directional ambiguity and sensitivity fading can be overcome by a phase modulation of the light source. If the branches are not identical or the splitting or combining of the beam is imperfect, then the MZI system has to be calibrated and an offset value measuring the blank determined: $\Delta\varphi_m = \Delta\varphi + \Delta\varphi_0$ with $\Delta\varphi_m$ as measured value and $\Delta\varphi_0$ as offset value.

With a photodetector allowing lateral resolution, e.g., a CCD camera, an interferometric approach based on the Young double-slit experiment (Thomas Young, 1773–1829) can be carried out. In the *Young-Interferometer* system, the sensing beam and the reference beam being identical abandon the sensor chip into the free space [11, 20, 35–37]. Sample and reference media are applied to the corresponding branches. Another approach buries the reference waveguide which excludes contact to a reference medium [38, 39]. In both approaches, the beams are diffracted at a double slit in the near-field or the diameter dimension of the diverging exit beams is so small that the exit plane acts as a double slit and they can interfere in the far-field. A CCD detector records the interference pattern.

The spatial power distribution $P(x)$ is given by

$$P(x) = \frac{1}{2}A(x)\left[1 + \cos(\frac{2\pi x}{p} + \Delta\varphi + \Delta\varphi_0)\right], \tag{46}$$

where x is the lateral coordinate of the interference pattern, $A(x)$ the envelope function deriving from the aperture of the waveguide, p the period of the interferogram, $\Delta\varphi$ the phase shift, and $\Delta\varphi_0$ an offset value of the phase shift (Fig. 13).

The period p is determined by the distance D from the double slit to the detector and the distance d between the two slits:

$$p = \frac{\lambda D}{d}. \tag{47}$$

For $D >> d$, the envelope function is determined by the diffraction at the single slits. The width of the interferogram B is defined by the distance of the two first order minima on each side of the maximum as

$$B = \frac{2\lambda D}{b} \tag{48}$$

with b denoting the slit width. For an exact recording of the interference pattern, a conventional CCD array with a pixel width considerably smaller than its period can be used. The close distance between the double slit leads to fringes of a period that can be detected without any magnifying lenses. The interferogram can be exploited

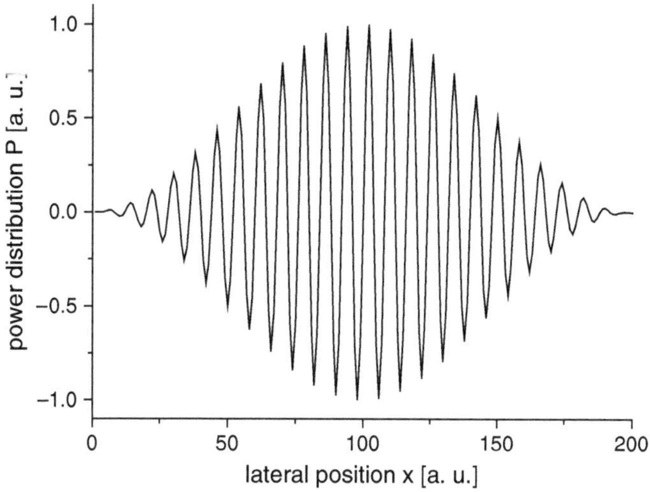

Fig. 13 Power distribution *P* as a function of the lateral coordinate. A phase shift in the wave representing the sensing branch causes a lateral shift of the interference pattern on the CCD detector

by a Fast Fourier Transformation process to obtain the phase shift induced by the sample on the sensing branch and the change of the effective refractive index. The lateral shift Δx of the pattern corresponds to the phase shift $\Delta\varphi$ by

$$\Delta\varphi = \frac{2\pi}{\lambda}\frac{d}{D}\Delta x. \tag{49}$$

A typical value of the refractive index is assumed so that the layer thickness or mass coverage on the waveguide can be calculated. If the polarization of the light source can be switched between TE and TM, then the refractive index and layer thickness or mass coverage can be determined unambiguously. For each of the two polarization states, the effective refractive index can satisfy a continuous range of thickness and refractive index values. But there is only one combination of values that satisfies the effective refractive index of both the TE and TM modes, simultaneously [38]. The mass coverage $\Delta\Gamma$ and the refractive index change Δn_{ad} can be separated as well by using dual wavelength propagating simultaneously through the waveguide or two different waveguide thicknesses [40]

$$\Delta\varphi = \frac{\partial\varphi}{\partial\Gamma}\Delta\Gamma + \frac{\partial\varphi}{\partial n_{\mathrm{ad}}}\Delta n_{\mathrm{ad}}. \tag{50}$$

Since the incremental changes of the mass coverage and the refractive index are different depending on the wavelength and the waveguide thickness, there are two equations to determine the unknown quantities $\Delta\Gamma$ and Δn_{ad}.

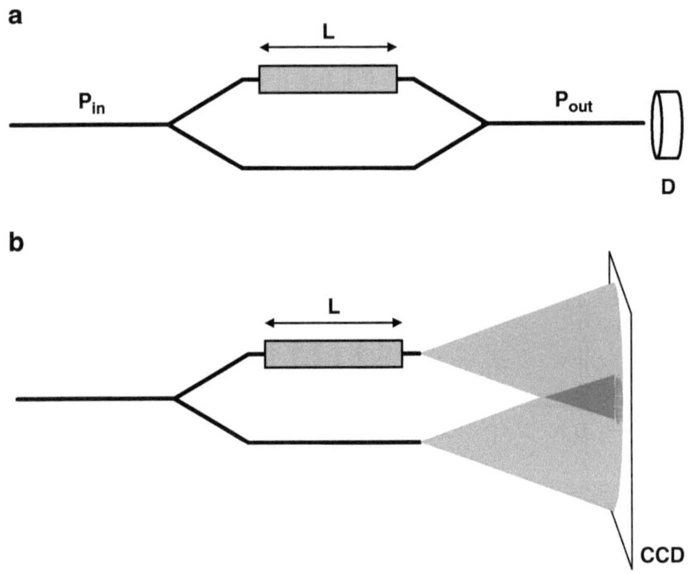

Fig. 14 Schematic drawings of interferometric arrangements: (**a**) Mach-Zehnder configuration: the sensing and the reference branch are combined after the interaction of the sample and a photodiode (D) that records the light intensity time-resolved. (**b**) Young interferometer. The sensing and the reference branch interfere in the far field. The interference pattern is recorded by a CCD. Either the end face of the waveguide structure approximates two point light sources or a double slit can be set in the beam path to generate the interference pattern. L interaction length, P_{in} input power, P_{out} output power

The Young interferometer overcomes the problem of sensitivity fading in the extremum areas since it resolves the signal laterally using a CCD sensor, while the photodetector of a MZI counts the fringes in a time-resolved manner (Fig. 14). But the Young interferometer approach has the same problem of ambiguity as the MZI, unless a semicoherent (or incoherent) light source is used. In this case, the interference pattern is superimposed on the envelope function known from a single slit. With an algorithm resolving the peak of the interference pattern only, the absolute phase value can be determined.

Compared to grating coupler sensors, interferometric systems are more sensitive and exhibit a very high resolution due to the long interaction length (typically in the range of 1–2 cm) of the guided mode with the sample and the high accuracy the phase shift can be measured (Tables 1 and 2). However, this high resolution cannot be exploited in practical biosensing applications since the sensor responds simultaneously to ambient or sample temperature fluctuations or refractive index changes of the sample solution. The contributions of the biosensing, refractometric, and temperature effects to the phase shift are superimposed and cannot easily be distinguished.

Table 1 Different detection limits of grating coupler sensors

Type	Waveguide material	Citation	Detection limit ($\Delta n_{\mathrm{eff,\ min}}$)	Detection limit by model system
Input	SiO_2–TiO_2	[26]	5×10^{-6}	6 μg/mL anti-h-IgG to adsorbed h-IgG
	SiO_2–TiO_2	[25]	2×10^{-6}	0.16 μg/mL anti-h-IgG to adsorbed h-IgG
	SiO_2–TiO_2	[55]		15 μg/L terbutryn (pesticide) in a competitive assay
	SiO_2–TiO_2	[56]		10–150 μg/mL mouse IgG to immobilized anti-mouse IgG 0.8–7 μg/mL mouse IgG in competitive assay format
	Ta_2O_5	[57]		$M_r = 2{,}000$ Da (biotinylated somatostatin)
Output	SiO_2–TiO_2	[27]	3×10^{-6}	4 μg/mL anti-h-IgG to adsorbed h-IgG
Reflected mode	Ta_2O_5	[30]	3×10^{-6}	0.2 μg/mL; 1.4×10^{-9} M h-IgG (to adsorbed protein G)
	Ta_2O_5	[18]	3×10^{-6}	10 pg/mm^2 BSA-biotin/streptavidin-multilayer, 0.25 μg/L simazine (herbicide) inhibition assay
Input and output	Ta_2O_5	[19]	$<10^{-6}$	0.3 pg/mm^2 biotin to immobilized neutravidin
Colorimetric resonant reflection	Si_3N_4[a]	[31]		3.4×10^{-5} Δn (determined by solvents with defined refractive index); 0.4 pg/mm^2 dried BSA; ~0.1–1.0 ng/mL streptavidin to immobilized biotin

[a]Also Ta_2O_5 and TiO_2 in other publications used

Table 2 Different detection limits of various interferometric systems

Type	Waveguide material	Citation	Detection limit ($\Delta n_{\mathrm{eff,\ min}}$)	Detection limit by model system
Mach-Zehnder	Si_3N_4	[58]	4×10^{-7}	2×10^{-5} Δn (glucose solution)
Young	SiO_xN_y	[11]	9×10^{-8}	750 fg/mL; 130 pM (50 ng/mL IgG to adsorbed protein G)
	Si_3N_4	[37]	2×10^{-8}	Glucose solution
	Ta_2O_5	[20]	9×10^{-9}	13 fg/mm^2; 390 fM IgG (to adsorbed protein G)

6.3 Evanescent Field Fluorescence

The utilization of fluorescence dyes for analytical measurements enhances the sensitivity for the detection of the molecules of interest. First, Cronick and Little made use of evanescent wave excitation for a fluorescence immunoassay, in 1975. By using totally internally reflected light, they excited the fluorescence of a fluorescein-labeled antibody which has become bound to a hapten–protein conjugate adsorbed on a quartz-plate in an antibody solution [41]. Contrary to the label-free high-refractive-index sensors where the mass of the molecule of interest is

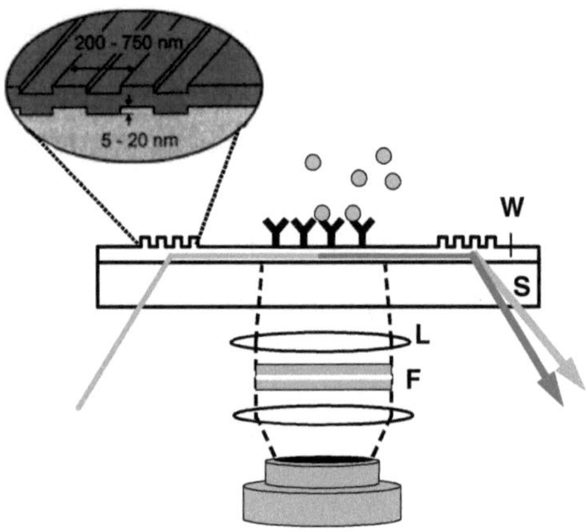

Fig. 15 Detection principle of evanescent-field fluorescence on planar waveguides: excitation light is coupled into a thin-film waveguide; surface confined fluorescence of bound labeled molecules is detected by a CCD camera. The fluorescence could be also separated via the outcoupling grating and detected by a photodiode. S: substrate, W: waveguide, L: lens, F: filter.

crucial, the sensitivity of the evanescent field fluorescence depends, in addition to the number of capture sites and affinity constants, especially, on the fluorescence dye. The higher the molar extinction coefficient and the number of fluorescent reporter molecules, the more sensitive is the fluorescence detection system. This is common to conventional fluorescence imaging systems. Yet, the illumination for the fluorescence excitation is limited to the thin evanescent field above the waveguide surface. Ta_2O_5 or SiO_2–TiO_2 are typical high-refractive-index compounds for this purpose [42]. The optical arrangement consists of beam forming optics for light coupling, including the interference filters for the excitation wavelength, and detection optics, including an emission filter to focus the image onto a CCD chip similar to a microscope or a scanning unit with a photodetector (Fig. 15). The evaluation of the fluorescence image is identical to the procedure used in biochip processing. Since the fluorescence excitation is confined to the evanescent field above the waveguide, the assay can be carried out in the presence of the sample. Furthermore, this method is suitable to immobilize the molecule or object of interest, e.g., a biological membrane fragment onto the surface and apply various potential binding partners to the probe. Many different sample solutions independent of pH or solvent can be conveyed sequentially. Independent from biochip or microscopic applications, the signal-to-noise ratio is enhanced by a factor of 100 compared to conventional epifluorescence [42–44]. The reported limit of detection is a spotted concentration of 2 pM Cy5-labeled immunoglobulin G per 150-μm-diameter-spot, corresponding to an amount of 0.8 zeptomole antibody or 500 protein molecules at a dye/protein ratio of ~5. The performance of this system was demonstrated by multiple marker quantification. A set of three cytokines, human interleukins IL-2, IL-4, and IL-6, were simultaneously measured on an antibody array as multiplexed sandwich immunoassay equivalent to an enzyme-linked immunosorbent assay (ELISA). The determined limits of detection are in the range of 1–15 pg/mL for all three investigated interleukins with a variation of signals within one microarray of $<10\%$. Regarding the chip-to-chip uncertainty, the assay precision showed a coefficient of variation $\leq 15\%$ [44].

But also other materials with lower refractive indices enabling total internal reflectance like glass slides can be applied as transducer. Then, an expanded laser beam can be launched into the edge of a microscope slide, thus providing even fluorescence excitation within the evanescent field [45].

7 Commercial Sensor Systems

In recent years, biosensors based on waveguide technology have attracted increasing attention due to their versatility and ability for use in medium and high throughput applications. Some of them took the step into commercialization and have to prove of value for industrial applications.

Farfield Sensors Ltd (Crewe, UK) developed the so-called *Ana*Light® sensor system based on dual polarization interferometry applying planar silicon oxynitride waveguides (http://www.farfield-scientific.com). A He–Ne laser illuminates the end-face of slab waveguide stack consisting of the sensor waveguide and the reference waveguide buried beneath after passing a ferroelectric liquid crystal half plate that switches the state of polarization between TE and TM. Since the distance between the two waveguides is only a few hundred micrometers, the diffracting light beams at the end of the stack generate the Young interference pattern that is recorded by a high-resolution CCD camera. The image data are synchronized with the polarization state and processed by a spatial Fourier transformation analysis. The system measures the thickness of an adsorbing layer with a resolution of <0.01 nm corresponding to 0.1 pg/mm^2 [46]. The fluidic handling is based on a pump that conveys the sample to the sensing waveguide. An adaptation to the widespread standard microtiter plate format, such as the 96-well and the 384-well microplates, is not available and seems challenging since the single wells cannot be separately accessed. But a parallelization is imaginable since the physical principle itself does not require moving parts so that the optical readout could scan several sensor areas.

The Centre Suisse d'Electronique et de Microtechnique SA (CSEM) developed the WIOS (wavelength-interrogated optical sensor) system based on the grating coupling method for the real-time observation of biomolecular interactions (http://www.csem.ch). It is a small bench-top system with normally 4–8 parallel fluidic channels for the label-free determination of binding interactions. High-refractive-index waveguide gratings with different grating periods of 300–700 nm for the coupling areas (input and output pad) are used as optical transducer. The mass coverage can be detected to a lower detection limit of 300 fg/mm^2.

SRU Biosystems (Woburn, MA) applied the principle of the colorimetric resonant grating reflection in the so-called BIND^{TM} (Biomolecular Interaction Detection) system that allows parallel readout in 96- and 384-well microplate format (http://www.srubiosystems.com). A grating with a period shorter than the reflected wavelength range (550 nm) and a depth in the range of 170 nm is structured by a silicon master wafer on liquid epoxy resin. The silicon wafer was etched in a shape that can be used as a replication mold for the liquid resin that is squeezed between the wafer and a polymer foil. After curing by UV light, the solid grating structure is peeled away from the silicon master. TiO_2, a high-refractive-index dielectric material, is sputtered on the surface structure to obtain the final sensor surface. Both the molding replication and the sputter process can be carried out on a continuous plastic sheet of several hundred meters in length. The TiO_2 surface with an optical grating creates a very sharp optical reflection at a particular wavelength that can be accurately traced as biological molecules bind to the surface. The signal is measured as the shift of the peak wavelength value (ΔPWV). Using low-cost components, the sensor is able to resolve protein mass changes on the surface with a resolution less than 1 pg/mm^2 [47].

Corning Incorporated (http://www.corning.com) introduced with the Epic® system, a resonant waveguide grating (RWG) biosensor system, as a label-free high throughput screening platform for biochemical and cell-based drug discovery [22]. It is very similar to the colorimetric resonant grating reflection system by SRU Biosystems but applies Nb_2O_5 as waveguide material. The thin grating structure of ~75–100 nm thickness is deposited on a glass substrate that is integrated as the bottom in a 96- or 384-well microplate. The assay sensitivity is stated as 5 pg/mm^2. Cells that adhere to the sensor surface by the extracellular matrix were studied upon various stimulations. The change of the refractive index within the part of the cell that is entered by the evanescent wave is monitored and evaluated depending on the experimental approach [21].

Zeptosens (a division of Bayer AG, http://www.zeptosens.com/en/) developed the ZeptoREADER for microarray applications based on glass slides with a Ta_2O_5 waveguide for evanescent fluorescence excitation [48]. The system is designed to be applied for spatially resolved fluorescence emission analysis. The setup exhibits a surface-confined, very strong electromagnetic field for efficient and selective excitation of fluorescent labels on the surface. An incident laser beam (635, 532, and 492 nm optional) is coupled into a Ta_2O_5 waveguide on top of a transparent substrate, via a diffractive grating. The evanescent field with a penetration depth of about 200 nm into the adjacent medium ensures that only fluorophores close to the surface are excited, which leads to an improvement of the signal-to-noise ratio as well as a highly sensitive and precise measurement. The fluorescence emission light is collected by an objective and focused on a CCD camera enabling imaging of waveguide chip areas of about 5 mm × 7 mm, leading to microarray applications. A filter wheel with interference filters selects the emission wavelength of the applied fluorophores. For medium and high throughput applications, Zeptosens equipped the system with a stacker capable of holding waveguide chips or carriers adapted to the 96-well plate format arranging five chips. Furthermore, they can use a 96-well plate consisting of a base plate with the waveguide structure and a mounted plastic 96-well structure. The sample can be applied using manual or robotic multipipettors or by a flow-cell.

8 Applications

Optical biosensors based on high-refractive-index waveguides can be applied in biotechnological and pharmaceutical research for target analysis, drug discovery, drug development, and disease diagnostics. Therefore, highly parallel and sensitive measurement systems are needed for a reliable investigation of DNA and proteins. Due to their consumption of small sample volumes, sensor systems using array formats can achieve a low price per test with high specificity and sensitivity. Depending on the application, they can be not only integrated in automatic high throughput analysis but also used for highly sensitive experiments in the research laboratory. The systems exhibit accessibility for the add-on of flow cells or pipetting devices. High-refractive-index sensor platforms in various extensions were developed with the advantage of high sensitivity and measurement in presence of the sample without any rinsing. The spatially limited evanescent field at the surface can be applied for both surface-confined fluorescence excitation and label-free detection of refractive index changes [49].

Microarray technologies based on imaging detecting optics enable the parallel measurement of a huge number of different biomolecules. In pharmaceutical and medical research, microarrays are applied for gene screening to detect novel drug targets, for gene expression monitoring to identify disease-specific genes, and for the determination of toxic effects of drug candidates. Furthermore, gene mutations and single nucleotide polymorphisms can be detected. In case of DNA microarrays, the interaction is limited to polynucleotide sequences or chemically similar polymers. Proteins are considerably more complex but they offer more insight into the real nature of disease and drug action. While DNA can be amplified by polymerase chain reaction, a similar tool does not exist for proteins so that low detection limits are required. Protein microarrays address various interactions of ligand–receptor pairs, such as antibody–antigen, enzyme–substrate, cell membrane receptor–ligand, protein–protein/peptide or protein–DNA/RNA [48].

Similar to mRNA, most of the cell proteins (up to 90%) are believed to be present as low copy numbers [50]. Therefore, when no amplification is possible, the highly sensitive evanescent field fluorescence technology is a promising solution to reliably measure low abundance proteins enabling detection of molecules at the picomolar level [44]. Capture molecules are immobilized for sandwich-like assays consisting of a capture antibody, the molecule (e.g., protein, antibody) of interest, and a secondary reporter antibody. A multiplexed immunoassay of this kind was developed for simultaneous quantitative monitoring of cancer biomarkers in tissue extracts on the Zeptosens platform [51]. Another approach is the spotting of the cell lysate on a hydrophobized waveguide and the subsequent incubation with a specific antibody and a secondary reporter antibody, e.g., for the investigation of tumor markers by spots of lysates of tumor and normal tissue and incubation with tumor-specific antibodies [52]. If an evanescent field fluorescence system is furnished with a suitable microscope objective, the distribution of any dye labeling a reporter antibody or similar molecules in a sample, e.g., cell membrane, can be investigated.

The feasibility was demonstrated on human fibroblast cells fixed and stained for vinculin, a focal adhesion protein [42].

In pharmaceutical research, the determination of affinity constants is a crucial step in the development of antibodies and also for affinity studies of new target molecules. Surface plasmon resonance systems are a widespread technology applied for this purpose [53, 54]. Waveguide sensors provide also an excellent tool for highly sensitive label-free detection of biomolecules. Various types of grating coupler systems and interferometric systems are suitable for the investigation of binding events in a label-free manner. On the basis of an antibody–antigen interaction, the determination of binding constants was shown [35]. Furthermore, chemical environmental pollutants such as herbicides or fungicides, as small molecules, can be detected at very low concentrations.

Generally, the application of a sensor system is determined by the chemical or biological recognition elements. They have to fulfill the following requirements:

Sensitivity: The limit of detection for a certain compound is determined by its affinity constant or binding constant to the interaction partner immobilized to the surface. Therefore, the choice of the capture compound affects remarkably the assay performance. Because of the opportunity to create antibodies against different kind of compounds, e.g., peptides, proteins, enzymes, vitamins, environmental pollutants, herbicides, fungicides, etc., they are frequently used as a versatile tool for sensor assay developments.

Specificity: An important criterion is also how specifically the capture molecule binds its interaction partner molecule. Nonspecific binding of matrix molecules to the sensing layer leads to false positive signals causing an error in the measurement values. Control experiments with compounds that show nonspecific adsorption, e.g., bovine serum albumin (BSA), are recommended. Blocking of the surface, meaning that a layer of e.g., BSA prevents the nonspecific adsorption of other matrix molecules and is displaced by the molecule interacting specifically, can also enhance the specificity in an assay.

Stability: Often, a regeneration of the sensor surface is desirable so that several samples can be measured with the same sensor surface. Therefore, regeneration solutions for stringent washing of the surface are applied, e.g., solutions with low pH or a considerable amount of detergents. The comparability between the assays performed with the identical sensor surface requires a sufficient stability against these chemical conditions.

A further issue is the steric hindrance: the interaction properties in a homogeneous assay are not directly transferable to the binding conditions at a surface/interface. The binding site of an immobilized capture molecule is accessible only from one side. It must be exposed on the surface in an optimal manner. Likewise, the capturing partner, e.g., peptide or protein, should be oriented with its binding-active domain toward the sample solution.

High-refractive-index platforms have a very broad application range, comprising microarray technologies, assays in microwell plate format, and binding studies in a flow cell if the aforementioned issues are taken into account.

9 Concluding Remarks

Optical chemical and biosensors have proven to be very sensitive, stable, and reliable detection systems, in the recent years. The sensor platforms based on high-refractive-index waveguides can be applied successfully for both label-free and fluorescence-based detection. Label-free techniques abate the uncertainty induced by the effect of the label on molecular conformation, inaccessibility of active binding sites due to the label, steric hindrance, or the inability to find an appropriate label that functions equivalently for all molecules in the experiment. Yet, the drawback of nonspecific binding retards the implementation of the label-free techniques as standard instrumentation for high throughput screening. Pharmaceutical screening involves the detection of small molecules as drug candidates where the limit of detection is important and the detection of proteins where the prevention of nonspecific binding is crucial. Even though some suppliers, e.g., SRU Biosystems or Corning Inc., offer technical solutions for this purpose, label-free sensors are not yet routinely applied for binding studies in a highly parallel format. The high costs for the fabrication of the optical transducers play a key role for the implementation in screening applications. However, label-free sensors are perfectly suited for the determination of affinity constants since the binding reaction is not disturbed by any label. The utilization of the evanescent field fluorescence techniques improved the limit of detection for microarray applications considerably and simplified the sample preparation because of the confined illumination of the dye-labeled molecules at the sensor surface. Targeted development for specific applications, e.g., point-of-care devices or diagnostics, will lead to a broader distribution of high-refractive-index sensor platforms.

References

1. Baird CL, Myszka DG (2001) Current and emerging commercial optical biosensors. J Mol Recogn 14:261–268
2. Keusgen M (2002) Biosensors: new approaches in drug discovery. Naturwissenschaften 89:433–444
3. Nice EC, Catimel B (1999) Instrumental biosensors: new perspectives for the analysis of biomolecular interactions. BioEssays 21:339–352
4. Saleh BEA, Teich MC (2007) Fundamentals of photonics. Wiley, New York
5. Snyder AW, Love JD (1983) Optical waveguide theory. Chapman and Hall, London
6. Kogelnik H (1988) Theory of optical waveguides. Springer, New York, Berlin, Heidelberg
7. Tiefenthaler K, Lukosz W (1989) Sensitivity of grating couplers as integrated-optical chemical sensors. J Opt Soc Am B 6:209–220
8. Schmitt K (2006) A new waveguide interferometer for the label-free detection of biomolecules. PhD thesis, Université Louis Pasteur, Strasbourg
9. Hu H, Lu F, Chen F, Shi B-R, Wang K-M, Shen D-Y (2001) Monomode optical waveguide in lithium niobate formed by MeV Si+ ion implantation. J Appl Phys 89:5224–5226

10. Korishko YN, Fedorov VA, Feoktistova OY (2000) LiNbO3 Optical waveguide fabrication by high-temperature proton exchange. J Lightwave Technol 18:562–568
11. Brandenburg A, Krauter R, Künzel C, Stefan M, Schulte H (2000) Interferometric sensor for detection of surface-bound bioreactions. Appl Opt 39:6396–6405
12. Germann R, Salemink HWM, Beyeler R, Bona GL, Horst F, Massarek I, Offrein BJ (2000) Silicon oxynitride layers for optical waveguide applications. J Electrochem Soc 147:2237–2241
13. Gorecki C (2000) Optimization of plasma-deposited silicon oxinitride films for optical channel waveguides. Opt Lasers Eng 33:15–20
14. Wörhoff K, Lambeck PV, Driessen A (1999) Design, tolerance analysis, and fabrication of silicon oxynitride based planar optical waveguides for communication devices. J Lightwave Technol 17:1401–1407
15. Lukosz W (1991) Principles and sensitivities of integrated optical and surface plasmon sensors for direct affinity sensing and immunosensing. Biosens Bioelectron 6:215–225
16. Höök F, Vörös J, Rodahl M, Kurrat R, Böni P, Ramsden JJ, Textor M, Spencer ND, Tengvall P, Gold J, Kasemo B (2002) A comparative study of protein adsorption on titanium oxide surfaces using in situ ellipsometry, optical waveguide lightmode spectroscopy, and quartz crystal microbalance/dissipation. Colloids Surf B Biointerfaces 24:155–170
17. Heideman RG, Lambeck PV (1999) Remote opto-chemical sensing with extreme sensitivity: design, fabrication and performance of a pigtailed integrated optical phase-modulated Mach-Zehnder interferometer system. Sensors Actuators B Chem 61:100–127
18. Piehler J, Brandenburg A, Brecht A, Wagner E, Gauglitz G (1997) Characterization of grating couplers for affinity-based pesticide sensing. Appl Opt 36:6554–6562
19. Cottier K, Wiki M, Voirin G, Gao H, Kunz RE (2003) Label-free highly sensitive detection of (small) molecules by wavelength interrogation of integrated optical chips. Sensors Actuators B Chem 91:241–251
20. Schmitt K, Schirmer B, Hoffmann C, Brandenburg A, Meyrueis P (2007) Interferometric biosensor based on planar optical waveguide sensor chips for label-free detection of surface bound bioreactions. Biosens Bioelectron 22:2591–2597
21. Fang Y (2007) Non-invasive optical biosensor for probing cell signaling. Sensors 7:2316–2329
22. Fang Y, Ferrie AM, Fontaine NH, Yuen PK (2005) Characteristics of dynamic mass redistribution of epidermal growth factor receptor signaling in living cells measured with label-free optical biosensors. Anal Chem 77:5720–5725
23. Tamir T, Peng ST (1977) Analysis and design of grating couplers. Appl Phys 14:235–254
24. Lukosz W, Tiefenthaler K (1984) Directional switching in planar waveguides effected by adsorption-desorption processes. In: Proceedings of 2nd european conference on integrated optics, Florence, Italy
25. Nellen PM, Lukosz W (1990) Integrated optical input grating couplers as chemo- and immunosensors. Sensors Actuators B Chem 1:592–596
26. Nellen PM, Tiefenthaler K, Lukosz W (1988) Integrated optical input grating couplers as biochemical sensors. Sensors Actuators 15:285–295
27. Lukosz W, Clerc D, Nellen PM, Stamm C, Weiss P (1991) Output grating couplers on planar optical waveguides as direct immunosensors. Biosens Bioelectron 6:227–232
28. Lukosz W, Nellen PM, Stamm C, Weiss P (1990) Output grating couplers on planar waveguides as integrated optical chemical sensors. Sensors Actuators B Chem 1:585–588
29. Brandenburg A, Gombert A (1993) Grating couplers as chemical sensors: a new optical configuration. Sensors Actuators B Chem 17:35–40
30. Brandenburg A, Polzius R, Bier F, Bilitewski U, Wagner E (1996) Direct observation of affinity reactions by reflected-mode operation of integrated optical coupler. Sensors Actuators B Chem 30:55–59
31. Cunningham B, Li P, Lin B, Pepper J (2002) Colorimetric resonant reflection as a direct biochemical assay technique. Sensors Actuators B Chem 81:316–328
32. Cunningham B, Qiu J, Li P, Lin B (2002) Enhancing the surface sensitivity of colorimetric resonant optical biosensors. Sensors Actuators B Chem 6779:1–6

33. Li PY, Lin B, Gerstenmaier J, Cunningham BT (2004) A new method for label-free imaging of biomolecular interactions. Sensors Actuators B Chem 99:6–13
34. Weisser M, Tovar G, Mittler Neher S, Knoll W, Brosinger F, Freimuth H, Lacher M, Ehrfeld W (1999) Specific bio-recognition reactions observed with an integrated Mach-Zehnder interferometer. Biosens Bioelectron 14:405–411
35. Hoffmann C, Schmitt K, Brandenburg A, Hartmann S (2007) Rapid protein expression analysis with an interferometric biosensor for monitoring protein production. Anal Bioanal Chem 387:1921–1932
36. Schneider BH, Edwards J, Hartman N (1997) Hartman interferometer: versatile integrated optic sensor for label-free, real-time quantification of nucleic acids, proteins, and pathogens. Clin Chem 43:1757–1763
37. Ymeti A, Kanger JS, Wijn R, Lambeck PV, Greve J (2002) Development of a multichannel integrated interferometer immunosensor. Sensors Actuators B Chem 83:1–7
38. Cross GH, Reeves AA, Brand S, Popplewell JF, Peel LL, Swann MJ, Freeman NJ (2003) A new quantitative optical biosensor for protein characterisation. Biosens Bioelectron 19:383–390
39. Cross GH, Ren Y, Freeman NJ (1999) Young's fringes from vertically integrated slab waveguides: applications to humidity sensing. J Appl Phys 86:6483–6488
40. Stamm C, Dangel R, Lukosz W (1998) Biosensing with the integrated-optical difference interferometer: dual-wavelength operation. Opt Commun 153:347–359
41. Kronick MN, Little WA (1975) A new immunoassay based on fluorescence excitation by internal reflection spectroscopy. J Immunol Methods 8:235–240
42. Grandin HM, Staedler B, Textor M, Vörös J (2006) Waveguide excitation fluorescence microscopy: A new tool for sensing and imaging the biointerface. Biosens Bioelectron 21:1476–1482
43. Neuschäfer D, Budach W, Wanke C, Chibout S-D (2003) Evanescent resonator chips: a universal platform with superior sensitivity for fluorescence-based microarrays. Biosens Bioelectron 18:489–497
44. Pawlak M, Schick E, Bopp MA, Schneider MJ, Oroszlan P, Ehrat M (2002) Zeptosens'protein microarrays: A novel high performance microarray platform for low abundance protein analysis. Proteomics 2:383–393
45. Ligler FS, Sapsford KE, Golden JP, Shriver-Lake LC, Taitt CR, Dyer MA, Barone S, Myatt CJ (2007) The array biosensor: portable, automated systems. Anal Sci 23:5–10
46. Ronan G (2004) Doubling up: dual polarization interferometry determines protein structure and function. SPIE's *oe*magazine 17–20
47. Cunningham BT, Li P, Schulz S, Lin B, Baird C, Gerstenmaier J, Genick C, Wang F, Fine E, Laing L (2004) Label-free assays on the BIND system. J Biomol Screen 9:481–490
48. Ehrat M, Kresbach GM (2001) DNA and protein microarrays and their contributions to proteomics and genomics. Chimia 55:35–39
49. Schmitt K, Oehse K, Sulz G, Hoffmann C (2008) Evanescent field sensors based on tantalum pentoxide waveguides – a review. Sensors 8:711–738
50. Miklos GL, Maleszka R (2001) Protein functions and biological contexts. Proteomics 1: 169–178
51. Weissenstein U, Schneider MJ, Pawlak M, Cicenas J, Eppenberger-Castori S, Oroszlan P, Ehret S, Geurts-Moespot A, Sweep FCGJ, Eppenberger U (2006) Protein chip based miniaturized assay for simultaneous quantitative monitoring of cancer biomarkers in tissue extracts. Proteomics 6:1427–1436
52. Templin MF, Stoll D, Pawlak M, Joos TO (2006) Protein microarrays: Neue Systeme für die Proteomforschung. GIT Labor-Fachzeitschrift 50:890–892
53. Homola J, Vaisocherová H, Dostálek J, Piliarik M (2005) Multi-analyte surface plasmon resonance biosensing. Methods 37:26–36
54. Rich RL, Myszka DG (2005) Survey of the year 2003 commercial optical biosensor literature. J Mol Recogn 18:1–39

55. Bier FF, Jockers R, Schmid RD (1994) Integrated optical immunosensor for s-triazine determination: regeneration, calibration and limitations. Analyst 119:437–441
56. Polzius R, Bier FF, Bilitewski U, Jäger V, Schmid RD (1993) On-line monitoring of monoclonal antibodies in animal cell culture using a grating coupler. Biotech Bioeng 42:1287–1292
57. Polzius R, Dießel E, Bier F, Bilitewski U (1997) Real-time observation of affinity reactions using grating couplers: determination of the detection limit and calculation of rate constants. Anal Biochem 248:269–276
58. Prieto F, Sepúlveda B, Calle A, Llobera A, Domínguez C, Lechuga LM (2003) Integrated Mach-Zehnder interferometer based on ARROW structures for biosensor applications. Sensors Actuators B Chem 92:151–158

Planar-Waveguide Interferometers for Chemical Sensing

Daniel P. Campbell

Abstract Interferometry is an optical technique that compares the differences experienced by two light beams traveling along similar paths. Planar waveguides have evanescent fields sensitive to changes in the index of refraction in the volume immediately above the waveguide surface. Placing a chemically sensitive film within this region provides the basis for chemical sensing. Film–analyte interactions change the index of refraction, causing the propagating light speed or phase to change in a direction of opposite sign to that of the index change. To measure this change, a reference propagating beam, which is adjacent to the sensing beam, is combined optically with the sensing beam, thus creating an interference pattern of alternating dark and light fringes. When chemical or physical changes occur in the sensing arm, the interference pattern shifts, producing a sinusoidal output. Waveguides and interferometers come in a variety of designs, but all rely on the evanescent field interacting with a chemically selective film to produce a measured response. The sensing mechanism can be passive (a physical change) or active (reactive sites in the film). Through a judicious choice of sensing films, interferometers can be designed to detect a wide variety of chemical and biological materials. Multi-interferometer devices with several different sensing films can be used to detect and identify a variety of different chemical or biological analytes either through specific sensing chemistry or through analysis of patterned response from an array of different films.

Keywords Interferometer · Planar waveguide · Thin films · Chemical sensing · Evanescent field

---

D.P. Campbell
Georgia Tech Research Institute, 925 Dalney Street, Atlanta, GA 30332-0810, Georgia
e-mail: Daniel.Campbell@gtri.gatech.edu

M. Zourob and A. Lakhtakia (eds.), *Optical Guided-wave Chemical and Biosensors I*, Springer Series on Chemical Sensors and Biosensors 7,
DOI 10.1007/978-3-540-88242-8_3,

Contents

Abbrreviations

FTIR	Fourier transform infrared spectroscopy
TM	Transverse magnetic
TE	Transverse electric
ARROW	Anti-resonant reflecting optical waveguide
Teflon AF®	Teflon amorphous fluoropolymer – Dupont trademark
HEPES	(4-(2-hydroxyethyl)-1-piperazineethanesulfonic acid) buffer
TCE	Trichloroethylene
DCE	1,2-dichloroethylene
VC	Vinyl chloride
TNT	2,4,6-trinitrotoluene
BTEB	Bis(trimethoxysilylethyl)benzene
IMTS	Iodotrimethylsilane
SAW	Surface acoustic wave
SWG	Super white glass
TATP	Triacetone triperoxide
TIR	Total internal reflection

Symbols

φ	Phase
n	Refractive index
n_c	Cover refractive index

n_s	Substrate refractive index
n_f	Waveguide film refractive index
n_{eff}	Effective mode index
n_o	Ordinary refractive index of birefringent material
λ	Wavelength
L	Pathlength
W_{eff}	Effective waveguide thickness
m	Mode number
z_c	Lateral phase shift in the cover
z_s	Lateral phase shift in the substrate
β	Propagation constant
Λ	Fringe period
Tg	Glass transition temperature
pKa	Acidity constant
dn/dT	Refractive index change with change in temperature
χ_c	Optical beam's penetration in the cover
χ_s	Optical beam's penetration in the substrate

1 Introduction

Optical interferometry is a technique used to measure various changes that may occur along an optical path. These changes may be the result of a change in the wavelength of the light, a change in the length of the path the light is traveling through, or a change in the refractive index of the medium through which the light is passing. Refractive index changes can occur from the reaction or interaction of a chemical species with the sensing medium, forming the basis for a chemical sensor. To measure this refractive index change, interferometry is used.

Interferometry takes two optical beams traveling along similar paths. One passes through the area where the chemical is present, while the other is isolated from the chemical that serves as a reference. Optically combining the two light beams produces an interference pattern, a series of dark and light fringes caused from the constructive and destructive interference of the two coherent light beams. While holding the pathlength and wavelength of the light used in the interferometer constant, any refractive index change occurring along the path of the guided beam results in a change in the phase or velocity of the light traveling through it. When the refractive index change increases, the phase or speed of the light slows down or decreases, and conversely, as the refractive index becomes lower, the speed or phase of the light increases. The amount of this phase change, $\Delta\varphi$, as a function of the change in refractive index, Δn, is expressed as

$$\Delta\varphi = 2\pi L\Delta n/\lambda, \tag{1}$$

where L is the length of the interaction and λ is the wavelength of the light used.

Interferometers come in many sizes and designs, but for their application in chemical or biological sensing, the design they usually take is a small waveguide device with a pathlength on the order of millimeter to centimeters. Planar waveguides have evanescent fields sensitive to the index-of-refraction changes occurring in the volume immediately above the waveguide surface. These evanescent fields represent only a fraction of the total optical field so that typically the amount of change in refractive index occurring in the sensing film produces a concomitant fractional change in the refractive index as measured by the interferometer.

The one feature common to all interferometers involves some means for interacting one optical beam with the chemical or biochemical sensing material on the waveguide surface with a second beam that serves as a reference. The combining of these two beams produce the interference pattern. Measuring the apparent shift in this pattern provides a measure of the change in refractive index occurring in either arm of the interferometer. With calibration, this shift yields the concentration of the analyte present. Two approaches commonly encountered include the Mach-Zehnder and Young interferometers. The interferometers that find use as chemical and biosensors usually are fabricated on a relatively small planar substrate such as those composed of planar waveguides. Interferometers rely on sensing chemistry applied to the sensing arm of the interferometer to concentrate, interact, or react with the chemical or biological analyte that one wishes to detect and measure. The sensing film provides a means of increasing the amount of refractive index change that occurs in the sensing arm of the interferometer and aids in identifying the compound or biological species present. Interferometers are not spectrometers. They provide no spectral information that would be helpful in chemical identification. However, through a judicious choice of the sensing film, the interferometer can provide information needed to identify the chemical present.

In interferometric biosensing, the bioreceptor, in the form of an antibody, aptamer, or single strand of DNA placed on the sensor surface, provides high specificity and affinity for its bio-conjugate antigen. In the case of biosensing, the placement of an antibody or a single strand of DNA, the interferometer can uniquely identify a species present; this is not due to any property of the interferometer but relies solely on the biological receptor's ability to interact with a specific organism, protein, or complementary DNA strand.

The relative small size of most waveguide interferometers permits the placement of several interferometers on a single substrate or the collection of several interferometers into a single system. Information obtained from an array of different sensing films on numerous interferometers provides the information necessary to successfully identify a single compound or to deconvolve the signals from a mixture of compounds through the use of pattern recognition and other chemometric approaches. The comparative simplicity of optical interferometers results in a low-cost, low-power, real-time sensing device.

2 History

Interferometry originates with the work of Thomas Young, who observed and reported an interference pattern resulting when coherent light passed through two closely placed slits and was projected on a screen. He reported his results to the Royal Society in 1803 [1]. The interference phenomenon he explored helped to establish the wave nature of light. For nearly two centuries, interferometry was relegated to making measurements through space or passing through a thin interaction volume, such as the Michelson interferometers used in Fourier Transform Infrared Spectroscopy, FTIR. With the development of optical telecommunication, interferometry was moved into a planar-waveguide structure. The idea of placing a sensing moiety on a planar-waveguide surface was first explored by Lukosz and Tiefenthaler [2]. These researchers placed a bioreceptor on the surface of a planar waveguide and used the exposed electromagnetic field above the waveguide to interact with a bio-conjugate reaction, thereby altering the phase of the light in the waveguide when the reaction occurred. During these first experiments, these researchers noticed that humidity changes affected the grating coupling used in these early biosensors. The humidity affected the coupling angles and efficiencies because of adsorption of and interaction with water vapor with the evanescent fields in these thin waveguide devices. Further experiments showed that these effects could be used as the basis for a sensor to monitor gas and humidity changes [3, 4].

Subsequent work using planar-waveguide interferometers was centered on exploring their use as biosensors. The use of planar waveguides for biosensors is reviewed in [5], and specifically for whole cell detection in [6]. An excellent review of waveguide and interferometric structures for sensor use has recently been presented [7]. On the whole, much less has been reported on the use of planar interferometers for chemical sensing, even though there are a wide variety of chemical interactions and reactions available. The use of planar-waveguide interferometers as chemical sensors holds great promise and interest owing to their low cost, low power, sensitivity, and their ability to be integrated into a multisensor device. This chapter explores the mode of operation of these devices, the various designs the devices can take, and the variety of chemistries, both passive and active, that find use in sensing a wide array of chemical compounds.

3 Planar Waveguides

Optical interferometers are readily fabricated and assembled using a wide variety of materials, techniques, and components that have arisen from the development of microelectronics, telecommunications, and optical computing. Fabrication techniques and instruments, including photolithography, ion etching, and material deposition to very controlled geometries and thicknesses, are well developed in the semiconductor industry. Devices such as diode lasers and megapixel array cameras provide inexpensive and high-quality sources and detectors for use in interferometric sensors. Computers on a chip and small, inexpensive liquid-crystal displays are commercially available. These take the images and signals produced by the interferometers and convert them chemometrically into displayed concentrations and identifications of the analytes being sensed, thereby yielding a complete sensor system.

Planar waveguides form the heart of most interferometric chemical sensors. The waveguide provides the conduit for the light used to interact with the sensing chemistry and provides the platform for the sensing film to be deposited. Although different optical designs have been explored, most interferometric techniques have comparable underlying sensitivities for a given interaction length prior to the deposition of the sensing film. The sensing film can use a variety of reactions and interactions to provide orders of magnitude increases in sensitivity for a given analyte. Generally, the sensors are composed of three basic components: the light source, typically a laser; the waveguide; and a detector.

Planar waveguides are related to fiber optics. Optical fibers have high refractive index cores surrounded by a lower refractive index cladding. Light is directed down the fiber through total internal reflections between the core and the cladding. The cladding confines the electromagnetic field associated with the guided beam, reducing loss to the surrounding environment. Fiber optics can be used in chemical sensing where the distal end of the fiber is derivatized with a sensing film. This configuration provides very little interaction between the light and the film. Sensing done in the cladding provides a greater interaction length where either the cladding serves as an adsorbing layer or a portion of the cladding is removed and replaced by a sensing film. Fibers have also been tapered down to increase the evanescent field interaction. However, most of the applications of fibers for sensing have used adsorption-based measurements through transmission or reflection or fluorescence-based measurements, none of which employs an interferometric scheme for sensing. A review of fiber-optic interferometry appeared in 1990 [8].

Planar waveguides provide the most obvious means of assembling a chemical sensor based on interferometry. Planar waveguides are composed of a high refractive index material positioned on top of a lower refractive index supporting substrate. The electromagnetic field that was confined by the cladding in the case of optical fibers now projects out from the planar-waveguide's surface. This field, called the evanescent field, is used to interrogate changes in the refractive index in the region directly above the waveguide surface. The planar configuration provides a support for placement and patterning of the sensing chemical films on the waveguide. These sensing films provide a means to concentrate the chemical compound of interest and, if possible, are made to react with the target compound though proper tailoring of the sensing chemistry. The incorporation of a reactive sensing group within the sensing film provides a means to greatly amplify the signal generated, by causing much larger phase changes in the sensing film than would occur from mean mass addition alone. In addition, the planar configuration of these waveguides not only allows easy deposition of sensing and reference films, but also permits the fabrication of various optical components onto or into the waveguide or the substrate. Gratings can be fashioned to couple light in and out of the device, and mirrors, beam splitters, and/or modulators can be used to direct and manipulate the optical beams.

Waveguides can be composed of any optically transparent materials including glassy material – SiO_2, SiO_xN_y, Si_3N_4, TiO_2, Ta_2O_5, and ion-exchangeable glass, among others – and polymers such as polystyrene and poly(methyl methacrylate). The sensitivity of the waveguide measured by the magnitude of the evanescent field that extends above the waveguide surface and that is available for interaction with the sensing film is dictated by the refractive index of the waveguide material, the thickness of the waveguide, and the refractive index of the substrate. Sensitivities increase typically with increasing refractive index difference between the waveguide and the substrate. Of the two major classes of planar waveguides – those with a graded refractive index and those where the refractive index steps up from the substrate to the waveguide – the step-indexed waveguides provide the greatest sensitivities. In addition to the inherent sensitivity of the waveguide arising from the composition of the substrate and waveguide materials, the chemical-sensing film deposited on the waveguide has the capability of drawing more of the evanescent field up into the sensing film, adding to the overall sensitivity of the waveguide interferometer. The resulting chemical changes occurring within the film have the ability to further amplify the resulting refractive index changes and thereby enhancing the sensor's ability to detect and measure the presence of a particular chemical compound.

4 Planar-Waveguide Operation

Planar waveguides are composed of thin films of a transparent dielectric material with a higher index of refraction than the substrate on which they are deposited. The light is confined within the waveguide through total internal reflection at the interfaces between the waveguide and the substrate and the waveguide and the cover. Optical guiding and the number of optical modes that are allowed to propagate must satisfy the transverse-resonance condition in order to exist. The transverse-resonance condition states that a guided beam must experience a 2π phase shift between equivalent points in the cycle of propagation and reflections in the ray trace translating along the waveguide. Figure 1 shows one of these cycles that represent the 2π phase shift between equivalent points (first and last ray arrows).

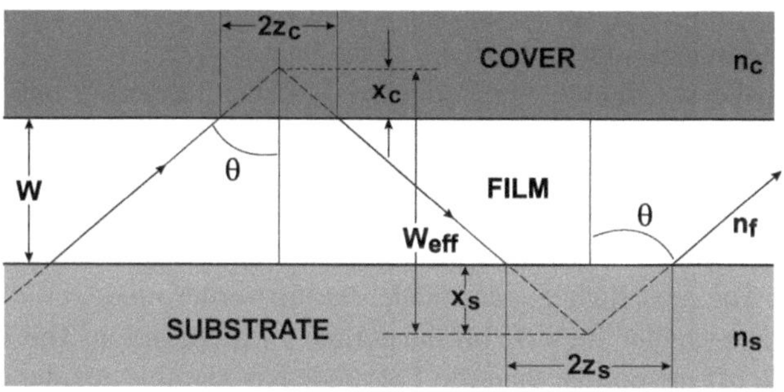

Fig. 1 Ray propagation in a waveguide

The following equations define the requirements for the transverse-electric (TE) and the transverse-magnetic (TM) modes of light propagation, respectively:

$$2\mathrm{k}n_{\mathrm{f}}\mathrm{W}\cos\ \theta - 2\tan^{-1}\left[\frac{\left(n_{\mathrm{f}}^{2}\sin^{2}\theta - n_{\mathrm{s}}^{2}\right)^{1/2}}{n_{\mathrm{f}}\cos\ \theta}\right] \\ -2\tan^{-1}\left[\frac{\left(n_{\mathrm{f}}^{2}\sin^{2}\theta - n_{\mathrm{c}}^{2}\right)^{1/2}}{n_{\mathrm{f}}\cos\ \theta}\right] = 2\ \pi m, \quad (2)$$

$$2kn_{\mathrm{f}}W\cos\ \theta - 2\tan^{-1}\left[\frac{n_{\mathrm{f}}^{2}\left(n_{\mathrm{f}}^{2}\sin^{2}\theta - n_{\mathrm{s}}^{2}\right)^{1/2}}{n_{\mathrm{s}}^{2}\,n_{\mathrm{f}}\cos\ \theta}\right] \\ -2\tan^{-1}\left[\frac{n_{\mathrm{f}}^{2}\left(n_{\mathrm{f}}^{2}\sin^{2}\theta - n_{\mathrm{c}}^{2}\right)^{1/2}}{n_{\mathrm{c}}^{2}\,n_{\mathrm{f}}\cos\ \theta}\right] = 2\ \pi m, \quad (3)$$

where m is the mode number (0,1, . . .), n_{f} is the waveguide's refractive index, n_{c} is the refractive index of the cover, n_{s} is the refractive index of the substrate, W is the thickness of the waveguide, λ is the wavelength, and $k = 2\pi/\lambda$.

Each of these equations accounts for the phase shift due to transmission through the waveguide medium, ($2kn_{\mathrm{f}}\ W\ \cos\ \theta$), and the two reflections, one off the waveguide-substrate interface (the second term in the equation), and the other off the waveguide-cover interface (the third term in the equation). Note that there are lateral shifts, $2z_{\mathrm{c}}$ and $2z_{\mathrm{s}}$, in the phase because of the total internal reflection of the surfaces. The shifts that occur at the interfaces represent the optical beam's penetrations into the cover and the substrate. This penetration, x_{c}, into the cover gives rise to the sensing evanescent field. Combined with the beam penetration, x_{s}, into the substrate, x_{c} adds to the thickness of the waveguide film, W, to give the effective thickness, W_{eff}, of the waveguide.

Note that in (3), an added multiplying factor appears in the equation for the TM mode. This results in the requirement of a thicker waveguide in order to support the TM mode. As the thickness required for TM guiding is achieved, the TE mode becomes more buried in the waveguide. Any of these modes can be used in interferometry; however, the somewhat greater extension of the TM mode's evanescent field provides added sensitivity to the final device.

As either the waveguide film's thickness or the refractive index increases, additional modes will satisfy the transverse-resonance condition, thereby allowing additional modes to be guided. The lowest-order mode, the zeroth-order mode, $m = 0$, is shown in the Fig. 2 ray model as the ray with the shallower propagation angle, even though the angle that is of importance is the angle θ, off the normal to propagation. The next higher-order mode, the first-order mode, $m = 1$, is shown propagating at a smaller angle, θ, off the normal for propagation. The TE mode has a larger angle off the normal than the TM mode has, but this angular difference is much smaller than the angular difference between the zeroth and first order modes.

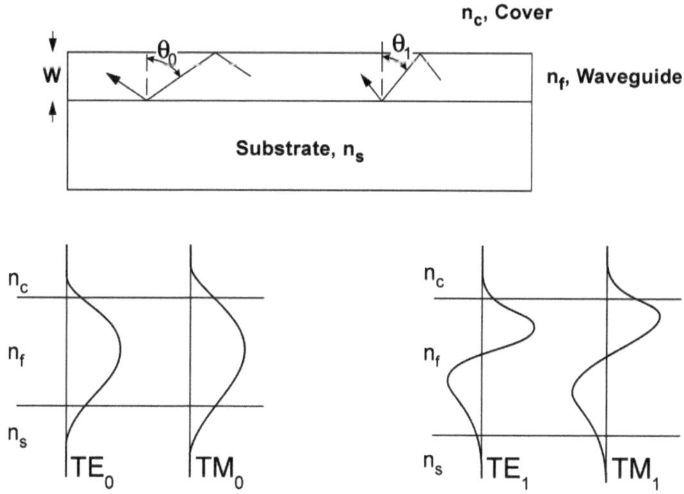

Fig. 2 Ray and field models for different modes of planar waveguides

As shown in the plot of the relative extent of the evanescent fields for both the zeroth- and first-order modes, it is suggested that using the zeroth-order mode would provide greater sensitivity for the waveguide system.

Waveguiding occurs only when the various parameters (the refractive indexes of the substrate, waveguide and cover layer; and the waveguide's thickness) are such that the transverse-resonance condition can be met. Waveguide behavior is presented in more detail in [9].

On examining Fig. 2's depiction of the field model, it is seen that for the higher-order modes, the evanescent field extends further above the surface. Also, the TM mode for a given order extends further into the cover than does the TE mode. Two interferometric designs take advantage of these relative differences in evanescent-field extensions to produce the interferometric signal: the polarimetric interferometer, which uses TE and TM modes of the same order to produce the interference signal; and the two-mode interferometer, which uses the difference between a lower-order mode and a higher-order mode to create the interference. In both cases, the reference arms of the interferometer provided by the TE and the zeroth order for the two types of interferometers described are collinear with the sensing arm. These built-in references are more confined to the waveguide, which helps minimize thermal and mechanical noise. They also provide cancellation of the background-refractive index changes since there is a small fraction of the evanescent field of the reference mode that is being affected by changes in the solution.

In the transverse-resonance conditions (2) and (3), only one term is affected by the chemical change occurring at the waveguide's surface. That term, n_c, reflects the change in the refractive index of the cover, and appears in the final term of the transverse-resonance equation. This contrasts with the free-space interferometry equation (1), where the phase change is directly proportional to the index of refraction in the medium through which the optical beam is passing. In the case of the waveguide interferometer, a smaller fraction of the total electromagnetic field is perturbed by the refractive index changes that are occurring only in the cover. As a result, (1) can be represented by (4) in the waveguide case, and reflects the effective change in the refractive index of the composite system, Δn_{eff}, i.e.,

$$\Delta\varphi = 2\,\pi L \Delta n_{\text{eff}}/\lambda. \tag{4}$$

The change in the effective mode index, Δn_{eff}, is the refractive index change that occurs in the waveguide's composite structure, a result of the change in the cover's refractive index. The value of n_{eff} is a function of the refraction indexes of the cover, waveguide, and substrate as well as of the thickness of the waveguide. The value of n_{eff} can be calculated by use of the transverse-resonance equation, using the angle of propagation θ, and the refractive index of the waveguide film, n_{f}, given as

$$n_{\text{eff}} = n_{\text{f}} \sin\ \theta. \tag{5}$$

The effective mode index differs from the refractive index of the waveguide film because the optical system is made up of contributions from the refractive indexes of the three or more layers that make up the waveguide structure. n_{eff} has a value that lies between the refractive index of the substrate and cover and the refractive index of the waveguide material. For example, a 1,500 Å layer of silicon nitride, $n = 1.85$, deposited on a fused silica substrate, $n = 1.457$, and having an aqueous solution serving as the cover, $n = 1.333$, has an effective mode index of approximately 1.56. When a biological organism, $n = 1.5$, is adsorbed to the waveguide surface, it displaces the water in that volume, increasing the cover's refractive index and in turn increasing the effective mode index, thus reducing the speed of light propagation in the waveguide which is equivalent to retarding the phase of the light. In chemical sensors, one may employ a sensing film comprised of a polymer which has a refractive index much greater than the 1.333 of water. This pulls a greater fraction of the evanescent field into the cover and when a refractive index change occurs within the film, a much greater change in the effective mode index results. Figure 3 plots the relative change of n_{eff} with increasing refractive index of the sensing cover film.

The greater the change Δn_{eff} produced by a chemical change occurring in the cover film on the waveguide, the greater the sensitivity of the waveguide interferometer. To produce a greater change Δn_{eff}, the waveguide system should be designed at the onset with a higher refractive index coating. This higher refractive index cover will pull more of the evanescent field up into the cover which has the effect of providing a greater contribution from the cover's refractive index to the effective mode index. A similar change in a cover will produce a greater interferometric response if the initial refractive index of the cover is higher to begin with.

There are two major types of planar waveguides – those with a graded index and those that are step-indexed, as shown in Fig. 4. Waveguides can be composed of any optically transparent material such as glassy materials and polymers, depending on their structure.

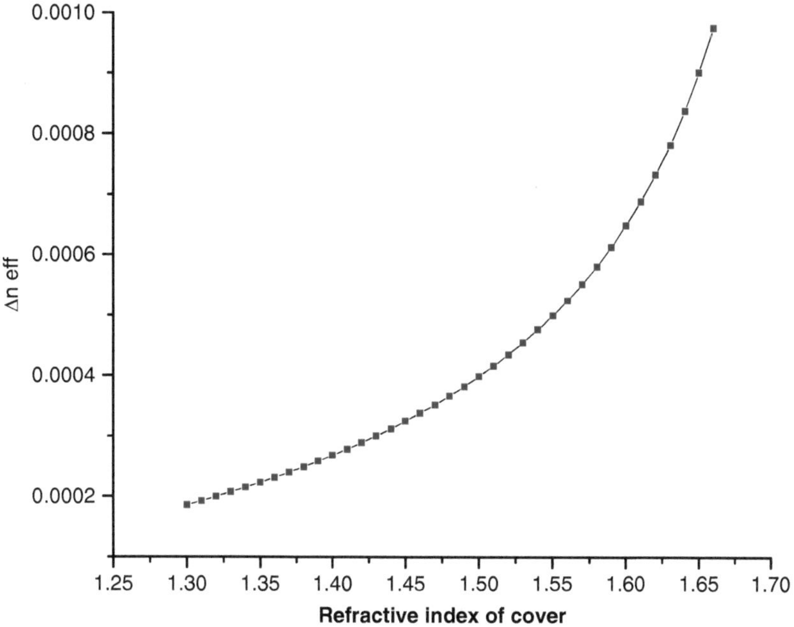

Fig. 3 Relative change in the effective mode index as a function of the refractive index of the cover

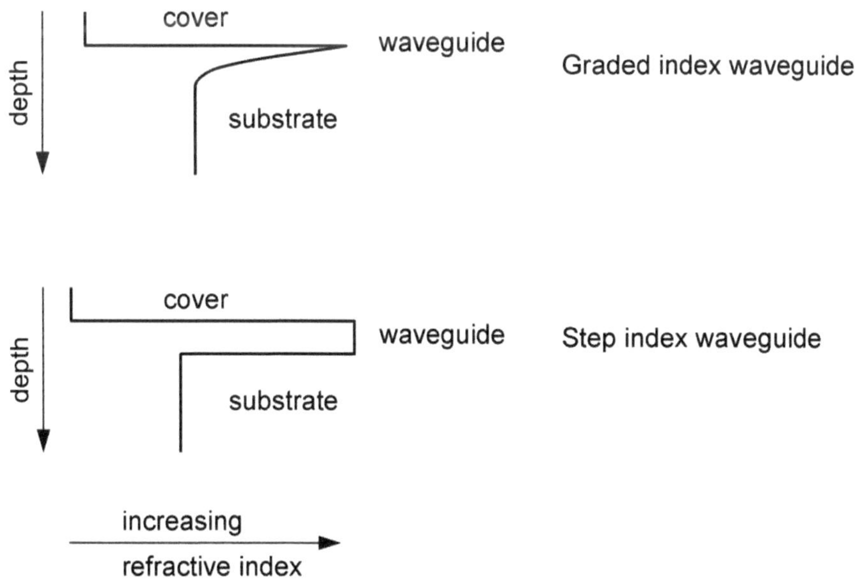

Fig. 4 Graded-index and step-index waveguide refractive index profiles

The graded-index waveguide has a refractive index that is high at the surface and gradually tapers down to the index of the substrate. This type of waveguide is produced using a substrate (usually glass) that has monovalent ions present, such as sodium, in its structure, that are exchangeable with silver, cesium, potassium, lithium, thallium, and other monovalent ions. The exchange is achieved by placing the substrate in a molten-salt bath that contains the ion to be exchanged. The concentration of the ion in the bath, the choice of that ion, the chemical makeup of the glass substrate, and the temperature of the salt bath will determine the resulting waveguide. The depth and amount of ion exchange in the glass will determine the refractive index gradient and the number of optical modes that the waveguide will support.

Further modification of the waveguide can be achieved by either applying an electric field while the waveguide material is in the molten-salt bath or by an annealing process after the bath. Index-of-refraction changes near the surface of 0.003–0.1 have been reported for graded-index waveguides [10–12].

Step-index waveguides show a step change in the index of refraction between the waveguide and the substrate. These waveguides are produced by depositing a high refractive index material onto a lower refractive index substrate. For polymer and sol-gel waveguides, the waveguide material can be spin-coated or dip-coated onto the substrate. For other glassy materials, processes borrowed from the semiconductor industry are used: chemical vapor deposition, ion-assisted deposition, plasma deposition, evaporation, and sputtering. These processes give fine control of the deposition process and allow one to tailor the properties desired in the waveguide.

The sensitivity is determined by the refractive index difference and the thickness of the waveguide. A list of waveguides and their relative sensitivities is shown in Table 1.

Evaluation of the sensitivity of various waveguide/substrate systems is easily accomplished by simply running a salt, sugar, or other known solvent or solute system that can alter the refractive index in a known or measureable way over the waveguide and measuring the phase change that results interferometrically. As the concentration of the salt solution increases, so does the refractive index of the

Table 1 Sensitivity of various waveguide systems

	Waveguide	Phase change per $\Delta n = 0.001$/cm	Relative sensitivity
Graded Index			
BK-7	0.25% Ag*/20 min	0.69 radians	1
BK-7	1%/10 min	1.35	2
SWG	0.25%/40 s	0.72	1
SWG	1%/8 s	1.13	1.7
Step Index			
SiO_xN_y (n = 1.85)	1,100 Å	23.9	34
SiO_2/SiO_xN_y	200 Å SiO_2 1,100 Å Si_3N_4	21.0	31
SiO_2/SiO_xN_y	400Å SiO_2 1,100 Å Si_3N_4	18.5	27.5
Ta_2O_5	1,200 Å	29.5	43

* % $AgNO_3$ in $NaNO_3$ @ 325°C

solution and, in turn, the effect on the guided mode in the waveguide. The graded-index waveguides are less sensitive than step-index waveguides because only a small perturbation of the index is possible using simple ion exchange. Step-index waveguides are best for maximizing the difference between the refractive indices of the waveguide and substrate. For example, an ion-exchange waveguide that was produced using a BK-7 glass substrate immersed in a molten-salt bath containing 0.25 mole% $AgNO_3$ in $NaNO_3$ for 20 min at 325 °C generated a 0.70-radian phase shift for a 0.001 change in refractive index in the cover solution over a 1 cm pathlength. A step-index waveguide made by chemical vapor deposition of 1,100Å of Si_3N_4 onto a fused-silica substrate produced a 23.9-radian phase shift for the same 0.001 change in refractive index above the waveguide over the same 1 cm pathlength. This 34-fold increase in interferometric sensitivity between the two waveguides points out the much greater magnitude of evanescent field associated with the step-index waveguide compared with the graded-index waveguide. The small evanescent field associated with the graded-index waveguides arises from the relatively small refractive index difference between the waveguide and the substrate and the required increased thickness of the waveguide in order to support waveguiding. For example, ion-exchange waveguides may have differences in refractive index of a few hundredths to thousandths. While step-index waveguides, such as the Si_3N_4 and Ta_20_5 waveguides, can have differences of 0.1–0.9 in refractive index between the waveguide and the substrate.

Figure 5 illustrates the Δn_{eff} of waveguide modes of different high refractive index materials for different modes and polarizations. The quantity Δn_{eff} is plotted

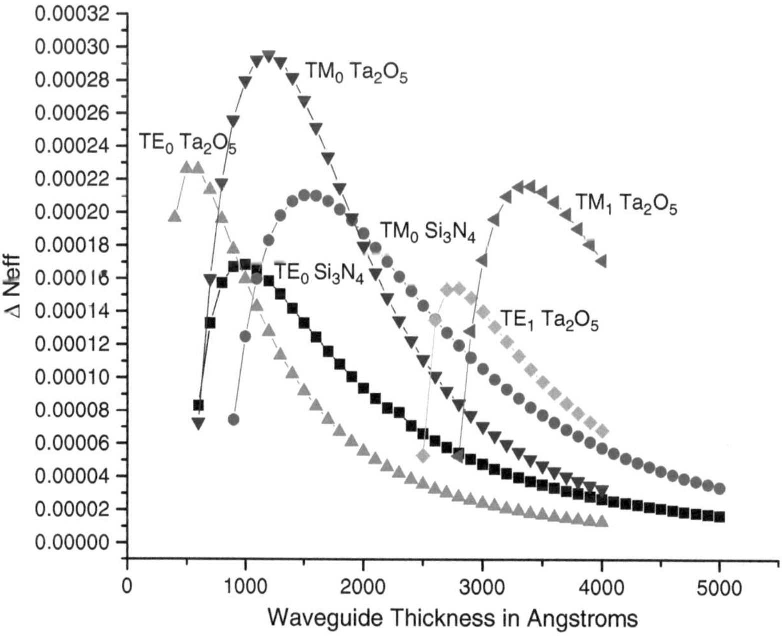

Fig. 5 Δn_{eff} for different waveguide systems, modes and polarizations

for a change in the cover index of 0.001 starting at the refractive index of water, n=1.333. It is quite apparent that TE modes propagate over narrow ranges and are more confined to the waveguide than TM modes. The higher refractive index materials such as Ta_2O_5 project more evanescent field up into the cover, topping out with 30% of the field residing in the cover. Also note that TE_0 and TM_0 offer more evanescent field than the higher-order modes TE_1 and TM_1.

Optimizing the refractive index difference between the waveguide and the substrate, in addition to decreasing the thickness of the waveguide to the point just slightly greater than the thickness required for waveguiding, results in a waveguide system where the evanescent field's extension into the cover is maximized and waveguide sensitivity is at its greatest. For chemical sensors, where the cover can be made of a wide variety of materials, increasing the cover's refractive index can produce a large effect on the sensitivity of the waveguide system. As the refractive index of the cover's refractive index is increased, more of the electric field becomes associated with the sensing film. It is even possible to have a waveguide material deposited that is not thick enough to guide until the sensing film is applied and produces conditions where waveguiding is permitted. This approach can lead to waveguide systems that are very sensitive to refractive index changes in the cover, since the cover is more closely coupled with the waveguide, thus producing very sensitive chemical sensors.

As previously mentioned, planar waveguides can be fabricated in either slab or channel designs. Slab waveguides have no three-dimensional structure but simply have a high refractive index film deposited over a wide area. In the slab waveguide, lateral modes become effectively infinite, eliminating the scattering possibilities between transverse and lateral modes. In channel-waveguide designs, Fig. 6, the interferometer use channels that are either embedded or ridged in structure. As the name implies, the embedded waveguides are produced by altering the refractive index of the substrate, usually by ion exchange. The channels are defined by the laser's light path or the sensing chemistry coating. To fabricate such structures a channel is defined photolithographically using a metal mask on an ion-exchange-able glass. The glass substrate is then immersed in a molten-salt bath where ions are exchanged in the open regions of the mask, producing a graded-index profile that is simply diffusion-driven and looks somewhat semicircular in profile, or the diffusion can be electric-field-assisted to produce more of a rectangular profile.

Alternatively, the channel can be photolithographically defined on the deposited high refractive index material and the material etched partly or completely away to give two versions of a rectangular-profile step-index waveguide. In addition, another type of channel waveguide has been used in which the lower refractive index channel is placed upon a thin layer of higher refractive index material, thereby forming the anti-resonant reflecting optical waveguide or ARROW which produces almost lossless propagation through the anti-phase reflection off the boundary planes of the high-index layer. In all the channel waveguides, the width of the channels has to be designed to avoid lateral modes in which light can be coupled into and interfere with the transverse modes. These effects are limited by proper control of the channel's thickness and the smoothness of the channel's sides.

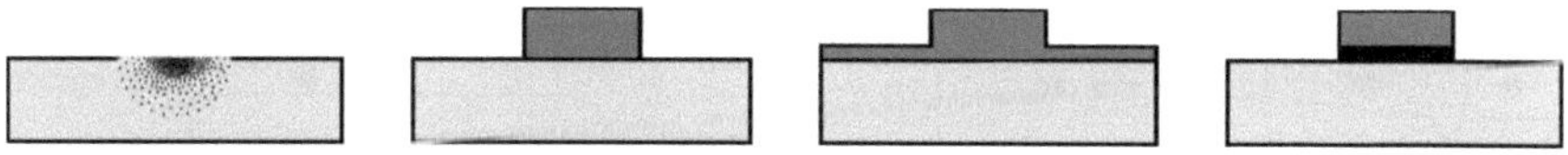

Fig. 6 Different types of channel waveguides – from left to right, ion-exchange, channel, ribbed, ARROW

5 Light-Coupling Methods

Waveguide interferometers consist of three components: a laser light source, a waveguide where the light interacts with the sensing chemistry, and a detector. Other optical components may be required to make connections between the various components, to manipulate the light before it is coupled into the waveguide, or at the output to direct the beams to interact or be directed to the detector. Interferometers can be constructed with varying degrees of integration ranging from an ensemble of discrete components to a monolithic device. One requirement for all interferometers is a means for getting the light into and out of the waveguide. Getting the light into the waveguide is not an easy task and several approaches exist. Coupling the light out is easier and may employ the same means as that used to launch light into the waveguide. The three means most commonly used for coupling the light into the waveguide include end-firing, prism coupling, and grating coupling, as shown in Fig. 7. Each method has advantages and disadvantages, depending on the design of the interferometric waveguide sensor.

End-firing is the most obvious method to excite a guided mode in a waveguide. The light can be either focused to a flat waveguide's edge by use of a lens or fed to this edge by the close positioning of an optical fiber. Maximum efficiency occurs when the incoming beam profile matches the guided beam profile [13]. For the zeroth-order mode, this profile is close to a simple Gaussian. The edge of the waveguide can either be cleaved, in silicon substrates, or polished to a flat edge to maximize coupling and minimize scattering. High coupling efficiencies are possible but not absolutely necessary, because the interferometric measurements are power independent and thus require only minimum intensities. End-firing works well when the waveguide is thick, such as for an ion-exchange waveguide. However, with a step-index waveguide in which the waveguide may only be a tenth of a micron thick, the positioning of the fiber becomes difficult. This precise positioning of the light source and waveguide becomes even more complicated when channel waveguides are involved, since the channel may be only a micron or so in width. These precise requirements in both horizontal and vertical positioning make end-firing a difficult method for coupling the light into a device, especially for one that requires frequent changes of the waveguide chips.

Prism coupling, in which a high refractive index prism is placed in intimate contact with the surface of the waveguide, permits highly efficient coupling of light into and out of a waveguide. Prisms such as heavy flint glass ($n = 2.009$) and rutile ($n_o = 2.584$) are commonly used. A bit of pressure and a scrupulously clean surface

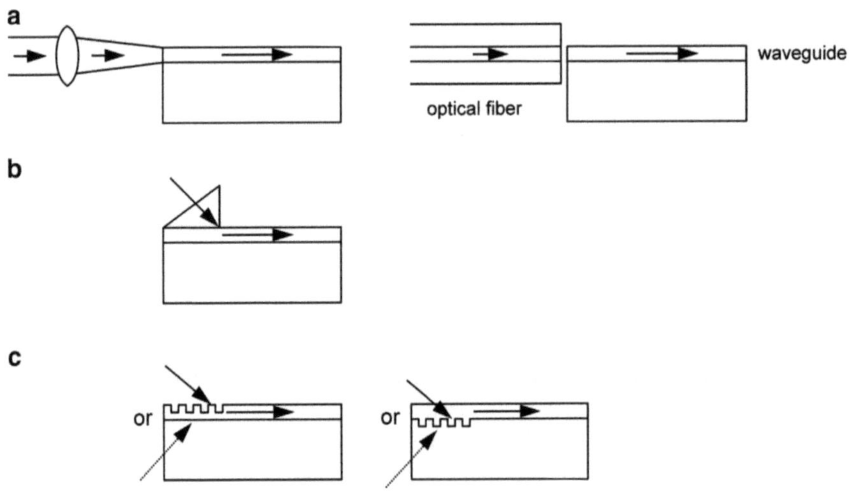

Fig. 7 Optical waveguide-coupling methods. (**a**) end-firing focused beam and end-firing optical fiber; (**b**) prism coupling; (**c**) grating coupling

are necessary for good contact. The visual appearance of a mirrored appearance at the interface is an indication that the required contact has been made. A slight beveling of the front edge of the prism provides the gap required for the light to jump from the prism into the waveguide, remain confined, and not couple back out upon reflection at the waveguide-substrate interface. Guiding occurs when the angle θ from the incident light in the prism equals the propagation constant

$$\beta = 2\,\pi\, n_{\mathrm{f}}/\lambda \sin\ \theta, \tag{6}$$

where n_{f} is the refractive index of the waveguide. If the waveguide is made to have additional modes, adjusting this angle allows one to address these different modes.

The use of prisms can find application when the sensing is conducted in air. However, when the measurement is taken with a liquid flow cell, the flow cell and the gasket between the flow cell and the waveguide become a major problem. The mechanism required to apply pressure to the prisms can also get in the way of the flow cell. The major problem involves the gasket used to seal the flow cell to the waveguide. The prisms need to be placed directly on the waveguide. The gasket around the flow cell will also be in contact with the waveguide at a point after the light is coupled into the waveguide. The gasket, if in contact with the waveguide, will decouple the light. To circumvent this problem, an isolation layer has to be fabricated between the prisms and the sensing area. This isolation layer allows light to travel under the gasket and into and out of the sensing area. However, the additional fabrication steps, the pressure-tightening mechanism on the topside of the waveguide, and the difficulty of changing out a waveguide chip, and replacing and realigning the prisms make this approach quite impractical for a commercial sensor.

It would be ideal to have a flow cell that can be placed on the waveguide with the light coming in from the other side. This brings us to another optical coupling device, the grating coupler. The grating coupler is a periodic structure with alternating refractive indexes that is fabricated into the substrate or waveguide by embossing, etching, or by ion exchange. A commercially available grating can be embossed into a deformable waveguide material, such as a sol-gel or polymer, to produce a grating. For example, the sol-gel SiO_2–TiO_2 is typically dip-coated onto a substrate and allowed to partially dry. A surface-relief grating is pressed against the coating with a pressure of 50–100 lb for a few minutes [2, 14, 15]. Upon release of the pressure, the grating die is removed and the sol-gel is cured thermally to produce the grating. Gratings from 0.25 to 0.85 micron, with efficiencies between 10% and 25%, have been formed using this method. But recall that interferometry is a power-independent measurement; so high efficiency is not imperative. Polymers can be used in a similar manner but they are less rigid, less inert, and have larger dn/dTs, which can alter the grating-coupling ability.

In an alternative procedure, the grating can be imaged photolithographically. The grating mask is first generated in one of a number of ways. One method uses two overlapping laser beams. A UV laser beam is divided into two beams with a beamsplitter. The two beams are initially spread out, spatially filtered, and then made to converge at a given angle to yield an interference pattern with the desired period. The patterning of the grating region can also be rendered through a shadow mask onto photoresist on the waveguide substrate or produced onto a master photomask. The resulting photoresist mask is then chemically developed and the grating is then etched with reactive ion etching or wet chemical etching [16]. The grating can be fabricated into the substrate or the waveguide. Any resulting gratings can be sealed by adding a protective film over them, such as SiO_2, to make them immune to changes in the sensing environment, while providing the refractive index contrast for the grating. These gratings have efficiencies similar to those of embossed gratings.

Gratings provide the option and advantage of having the light come in from the bottom of the waveguide, thus allowing for direct sensing of the environment, or the placement of the test cell and related fluidics on the sensing side. Gratings have less stringent alignment conditions than does the end-fire approach. Input optics can be designed to provide a cone of light with a reasonable fraction of the light impinging on the gratings at both the correct angle and position for coupling into the waveguide. The grating design allows waveguide chips to be routinely replaced, producing a "plug and play" sensor. Gratings work well for slab waveguides but pose a more difficult task for channel waveguides. The narrow width of gratings on a channel waveguide limit the amount of light that can be coupled in as well as making alignment a more difficult task. Gratings with large widths have been used to excite several interferometers on a single chip, at once, with uniform coupling efficiency [16, 17]. The choice of whether to input light into a waveguide using either a fiber to couple into the end of a waveguide or to couple a focused beam through a grating seems to be driven by the individual researcher's equipment and experience without any fundamental reason behind the choice.

6 Types of Interferometers

Planar-waveguide interferometers can be configured in a number of ways, but can be classified by their output. The two arms of an interferometer are combined either into a single collinear beam with infinite fringe period or at a slight angular configuration that yields a finite fringe period and an image that represents several interference fringe lines. Combining the two arms of the interferometer into a single collinear path produces a fringe pattern of infinite period and an output that represents only a point along this infinite fringe pattern. A phase shifter can be incorporated into one of the arms of the interferometer to allow for adjusting the phase or position on the fringe pattern to measure the minimum–maximum of the fringe or to place the interferometer at quadrature. Combining the two beams at an angle (even slight) will produce a defined fringe pattern that can be imaged in total on an array detector. The simplest method for producing a fringe pattern is to allow the separate arms of the interferometer to exit from the waveguide and then to be combined either through natural divergence or by the use of a lens, producing a multiple fringe pattern that can be imaged on a camera or CCD array. Total fringe monitoring eliminates any need for adjustment to quadrature, for the complete fringe provides all the information needed to measure and average the phase changes occurring on the interferometer. The period of the fringes, Λ, is determined by the spacing between the two channels and the distance to the point of convergence of the two beams; i.e.,

$$\Lambda = \lambda L/d, \tag{7}$$

where d is the distance between the two arms of the interferometer and L is the distance from the output to the overlap. The closer the spacing between the two arms of the interferometer and the length from output to overlap, the greater the period of the fringe spacing.

Four types of Mach-Zehnder interferometers are shown in Figs. 8a–d. The light for these interferometers is injected from a single point into the waveguide and then divided into the branched arms through the use of a Y-junction. The two arms serve as the sensing and reference arms of the interferometer. After passing through and under the sensing and reference arms, the two channels are brought together through use of another Y-junction which combines the two beams into a collinear output channel. Because the two beams are made collinear, there is no angular difference between the two beams. This produces an effective fringe pattern that is infinite in period. A single intensity measurement is made at the output, which corresponds to some point along the sinusoidal curve that a fringe pattern would have. Placing a phase shifter in one arm of the interferometer allows the interferometer's output phase to be adjusted. The phase shifter may consist of an electrooptic polymer or other electrooptic material or piezoelectric material. When a voltage is applied to it, the phase of the beam traveling in that arm of the interferometer is affected and alters the point on the sinusoidal fringe output of

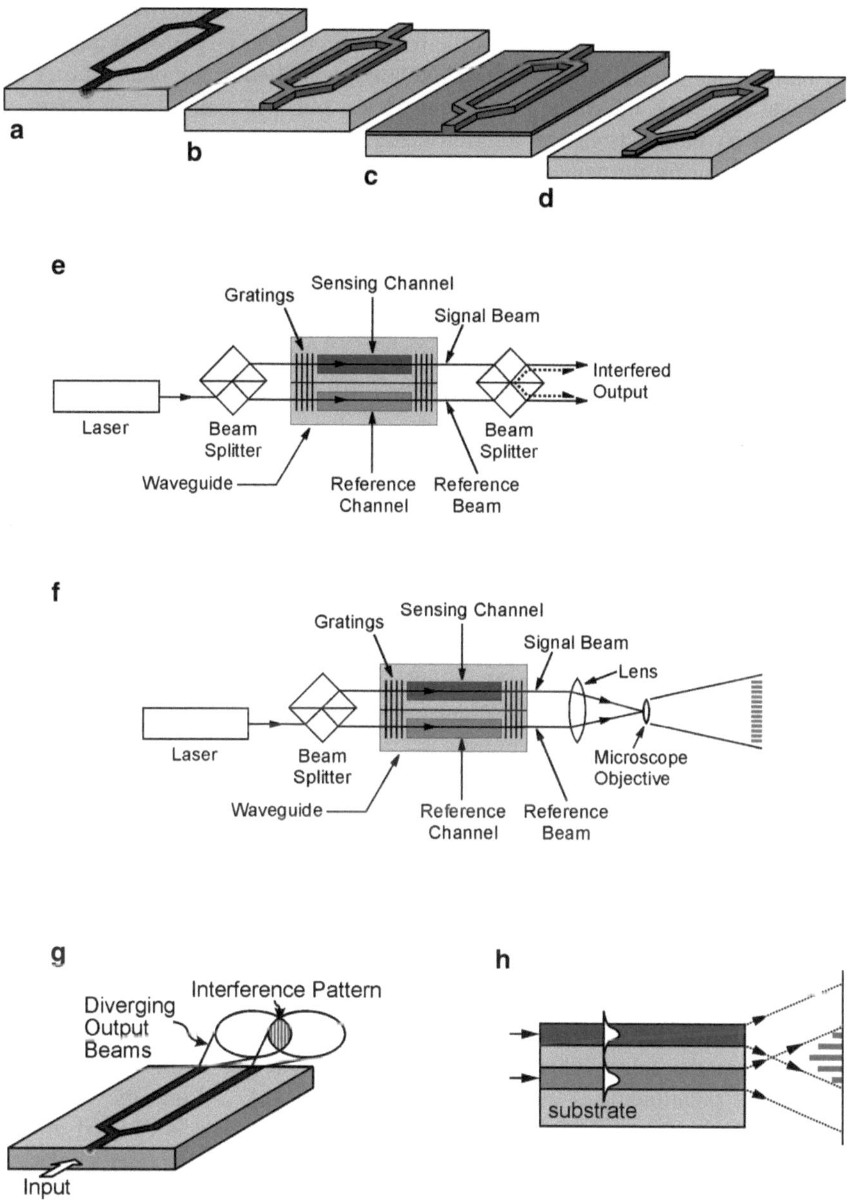

Fig. 8 Different types of interferometers – (**a–d**) Mach-Zehnder – (**a**) ion-exchange, (**b**) channelized, (**c**) ribbed, (**d**) ARROW, (**e–h**) Young – (**e**) using beam splitter for beam combining, (**f**) using lens for beam combining, (**g**, **h**) using dispersion for beam overlap in a side-by-side configuration, and (**h**) a sandwich configuration

the interferometer [18, 19]. Adjusting the phase allows the interferometer to be scanned through the min and max of the fringe and then to be placed at quadrature for optimal sensitivity. Adjustment of the voltage applied to the phase shifter not only can place the interferometer at quadrature but also will allow a means of providing an output from the interferometer. Rather than measuring the phase change, measuring the voltage required to hold the interferometer at quadrature provides the measure of interferometric response, and the chemical to be sensed can be calibrated to this applied voltage.

The Young interferometers, Figs 8e–g, can be very similar to the Mach-Zehnder interferometers. One can imagine the Mach-Zehnder design being cut off after the interaction area. The output of each channel is allowed to interact to form the interference pattern and the measurement of phase can be made directly from that interference pattern [20, 21]. This pattern can result from the natural divergence of the two output beams or the overlap can be forced to occur by the use of a simple lens placed at the interferometer's output [22]. Figures 8a–c, f show interferometers in which the light comes in at one point and is distributed into two channels. Figures 8e, f shows designs where the two beams are formed using a beamsplitter prior to coupling into the waveguide. The two approaches are coupled into the waveguide by means of a grating coupler. A slab waveguide serves as the two arms of the interferometer. The beams are coupled out using a grating coupler. If a beamsplitter is used to combine the output signals (Fig. 8e), a collinear set of outputs results with an infinite fringe pattern [23]. A phase shifter external to the waveguide allows adjustment of the phase position of the interferometer. If instead of a beamsplitter, a lens is used, a finite fringe pattern is produced (Fig. 8f). Since the two arms of the interferometer are a couple of millimeters apart, the resulting fringe spacing is very fine [22]. A microscope objective placed at the focus of the lens expands the interference to allow the pattern to be imaged directly onto a camera or CCD. Alternatively, placing a mirror on a pivoting stage after the output from the microscope objective allows the mechanical shifting of the array across a slit position at a distance such that the slit is only a fraction of the period in width. This permits the scanning of the fringe's min/max and setting at quadrature.

Other versions of interferometers that employ sensing and reference arms that travel within the same volume of waveguide and interact along the same path include the polarimetric or difference interferometer [24, 25] and the two-mode interferometer [16, 26]. In the polarimetric interferometer light is coupled in with both TM and TE polarizations. This is accomplished by rotating the laser's polarization to 45° from normal which will send a fraction of both polarizations of light into the waveguide. Owing to the differential extensions of the evanescent fields of the two polarizations, TE and TM light interact differently with the waveguide coating. TE light is more buried in the waveguide, and serves as a reference, while the TM light which extends further above the waveguide interacts to a greater extent with the sensing chemistry. At the output of the interferometer, a series of optical components are needed to rotate and interact the light from both polarizations with each other and provide the interferometric output.

The polarimetric or difference interferometer is shown in Fig. 9. An integrated version of the polarimetric interferometer has been designed to simplify the difference interferometer, shown in Fig. 10 [27]. The two-mode interferometer uses different modes within the same channel having different evanescent field penetration depths to differentially interrogate the sensing film, as shown in Fig. 11. This design requires additional optics for coupling in the light at the different angles required to excite both modes. Also, the outputs emerge at different angles requiring additional optical components to combine the two beams to produce the interferometric pattern.

In addition to the various designs described heretofore, some researchers are moving sensors into other directions. One group is working on integrating the various components of the interferometer and incorporating them all – source, interferometer, and detector – onto the same hybrid chip [28]. The advantage of this integrated device is the minimization of mechanical noise and the small size the entire device can achieve. Another group is working on an active sensor. This is not to be confused with the active sensing discussed further in this chapter. This device involves making the sensor active by electrooptically modulating both arms of the interferometer to remove the drift in input power, phase offset, and mechanical instabilities as well as thermal fluctuations. This frequency-modulated Mach-Zehnder has resolution capabilities for measuring refractive index changes on the order or 5×10^{-9} [29].

A couple of multiple-interferometer systems have been reported. A 13-interferometer chip has been fabricated on a 1×3 cm chip complete with grating couplers for getting the light in and out of the interferometer, beam splitters, and TIR-total reflective elements which are used to combine the beam and produce an infinite fringe pattern. A four interferometer portion of this chip is shown in Figure 13, a [26, 30]. Two multiple Young interferometer designs have been described. In Figure 12, a channelized waveguide has its light is divided again and again, then directed into three separate interferometers which have three different spacings between the sensing and reference arms [31, 32]. The three interferometric outputs produce three sets of interference patterns that overlap each other; however, due to the spacing differences between the sensing and reference arms, each interference pattern has a different spatial frequency. The close-spaced interferometer produces a wide fringe pattern, the medium-spaced interferometer produce a narrower pattern, and the broadly spaced channels produce a very finely spaced pattern. All patterns are imaged onto a single array of detectors. The separate signals are separated by their spatial frequency using the Fourier Transform.

The other multiple-interferometer chip design is shown in Fig. 14 [17]. This device has four pairs of interferometric channels of equal spacing but relies on a series of staggered output grating couplers to direct the output of each interferometer to sequential regions on the detector array. Current versions of this chip contain eight interferometers on a 1×3 cm glass substrate. Additional interferometers can be squeezed onto the same-sized chip. Alternatively, rather than staggering the gratings, the same series of interference patterns can be produced using a series of gratings with slight increases in the grating period. In this case, the outputs diverge from the different outputs at slightly different angles from the chip.

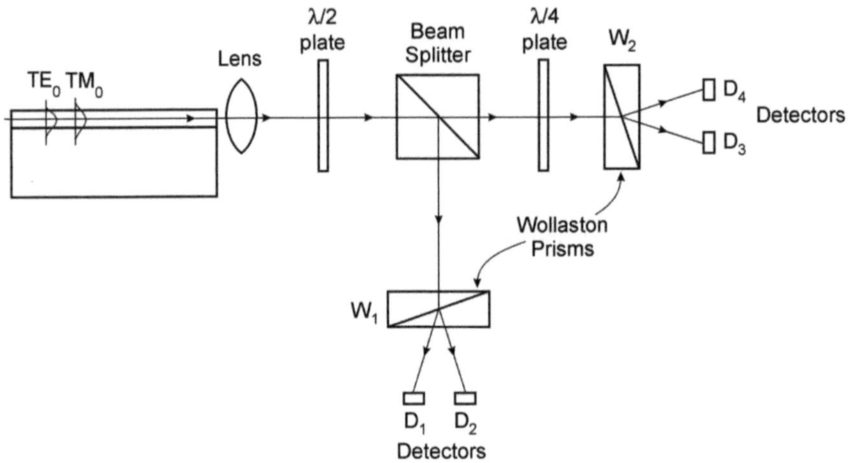

Fig. 9 Polarimetric or difference interferometer

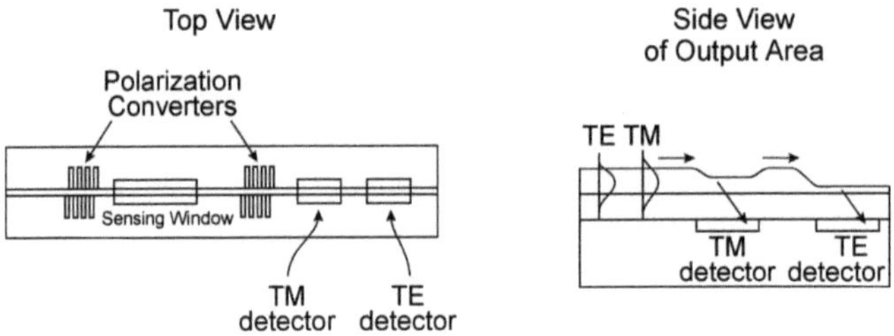

Fig. 10 An integrated version of the polarimetric interferometer

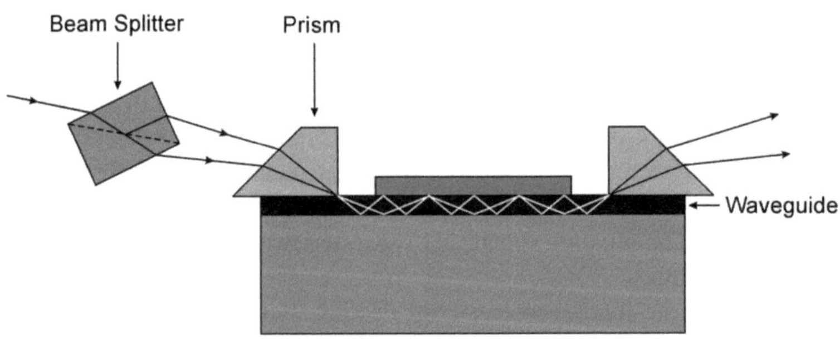

Fig. 11 Two-mode interferometer

Multiple-interferometer chips find use where different sensing chemistries are placed on each interferometer so that the interferometers react and measure singularly and specifically. Specific antibodies derivatized onto separate interferometers may be encountered. However, for most sensing applications, the sensing chemistry is not that specific and straightforward. Multiple interferometers will be needed

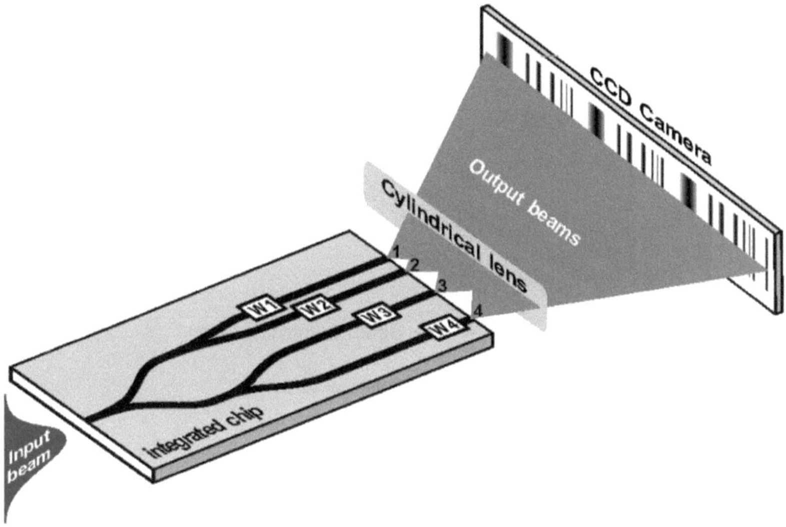

Fig. 12 Multi-interferometer chip employing different output spacing to produce varied fringe period

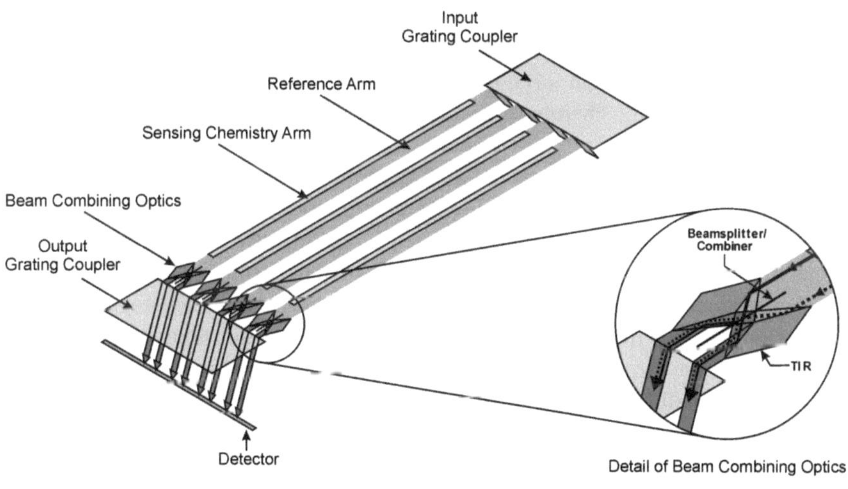

Fig. 13 Integrated multi-interferometer sensor with on-chip beam combining

since the compounds present may or may not be related in structure and reactivity. However, the compounds will have different indexes of refraction so that, by using an array of sensing films with similar sensing chemistries and different refractive indices, or with similar refractive indices but different reactive chemistries on the sensing channels, the determination of the composition and quantification of what is present is possible with proper calibration and chemometric algorithms.

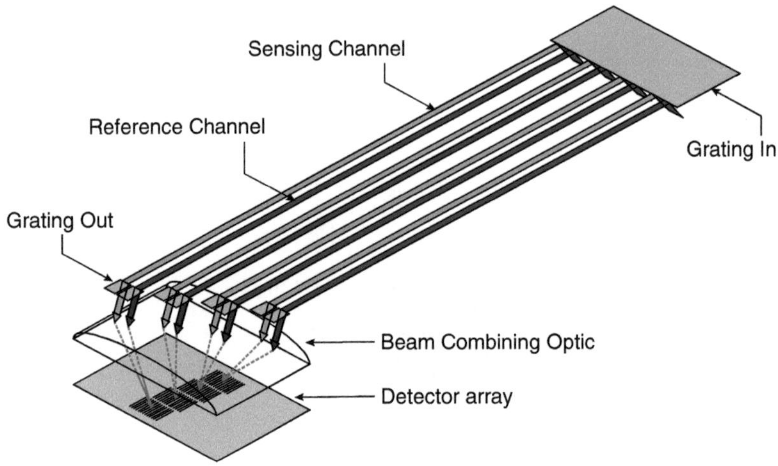

Fig. 14 Staggered output grating multi-interferometric waveguide sensor

7 Chemical Sensing – Passive and Active

The underlying sensitivity of any of the interferometers described heretofore is dictated by the waveguide's composition – substrate, waveguide, waveguide thickness, and interaction length – but the ultimate sensitivity of the sensor depends on how well the sensing chemistry can amplify the signature of any compounds present through adsorption or chemical reactions that alter the refractive index of the sensing film. The Δn_{eff} produced as a function of the waveguide's composition alone is typically a fraction, from a few percent to 30%, of the Δn occurring within the sensing medium. Different interferometric designs offer no inherent improvements from one design to another except in regard to their practical implementation. The greatest contributor to the change Δn_{eff} can occur from the sensing film and its interaction with the material being sensed. The sensing films applied typically are selected from the large array of polymers available, or from materials that can be designed and synthesized, or that can be doped with a reactive moiety. Other than polymers, sol-gels can also be used as sensing films that can be tailored for a given refractive index or reactivity.

Planar-waveguide interferometers respond to changes in refractive index occurring in the volume immediately above the surface. A chemical film placed on the waveguide surface can select and concentrate various analytes through a number of mechanisms that will produce measureable signals on the interferometer. Every chemical reaction and interaction can produce changes in refractive index. These interactions and reactions can be categorized as either passive or active in their mode of sensing. Passive mechanisms can occur alone or in combination with an active moiety added to the passive sensing film. Passive sensing involves physical changes in the film caused by adsorption and concentration of a chemical analyte from the surroundings causing the film to swell, filling in void space within the film, or dissolving into the sensing film. Even though a variety of various chemical factors come to play in determining the affinity a film has for a particular compound, it is just easier to put these factors aside and call the mechanism passive since no obvious reaction is occurring. Sensing films can concentrate the analyte present so that the sensor is not merely measuring the direct concentration of the analyte in air or solution but measuring material partitioned into the film. This concentrating effect can result in levels of an analyte in the film that is 10^5–10^6 times the concentration in the sensing medium. For example, the sensor rather than responding to a solution concentration of 10 parts-per-billion is responding to 1–10 parts per thousand being concentrated in the sensing film as determined by the partition coefficient, 10^5–10^6, respectively. These concentrating effects are driven by molecular similarities such as polarity among other effects and draw the material out of a water solution and into the organic sensing film with similar molecular composition (like dissolves like).

Active sensing mechanisms can change the refractive index to a large degree through the creation of a new compound within the sensing film and can provide enhanced responses to small amounts of material present. A combination of the two mechanisms, passive and active, allows adsorption-driven response by the partitioning to be followed by a chemical reaction within the film and further increases the interferometric signal produced, assuming that the refractive index changes occurring possess the same sign. The mixing and matching of different mechanisms allows a means to assist in the detection and identification of a particular analyte present in the sensing medium. And finally, a color change can enhance a interferometric signal. If the color change occurs upon adsorption of the analyte into the sensing film and some of the laser light is absorbed, there can be a change in the imaginary part of the complex refractive index (absorption), thus changing the signal.

In bio-sensing applications, mass addition rules the magnitude of the response that can occur in the interferometer. The binding of a biomolecule or bioorganism to a receptor bound to the waveguide surface with a concomitant displacement of any aqueous or buffered solution dictates the amount of response possible. Little if any contribution to the signal results from electronic changes occurring in the bio-conjugate reaction. The solution above the waveguide (water) produces interferometric responses that are limited by a waveguide composite structure having water as the cover. This also limits the evanescent field's extension into the cover. However, the substrate's refractive index can be subsequently lowered to values less than the refractive index of water through the use of fluorinated materials or highly porous sol-gels, although these applications are not that straightforward to put into practice. In chemical sensing, the sensing film can have a wide variation in refractive index, thereby allowing one to tailor the contribution of the cover to the composite waveguide structure; that is, more of the evanescent electromagnetic field can be drawn into the cover film and is, in turn, perturbed to a much greater extent by any chemical reaction or interaction.

7.1 Passive Sensing

Passive-sensing mechanisms are employed when the compounds to be detected tend to be unreactive, such as aromatic and chlorinated solvents. These compounds have had widespread use and due to their toxic nature require sensing and monitoring in groundwater applications, in drinking water, and in the air. Passive-detection chemistry relies upon the film, such as a polymer, adsorbing the analyte and physically changing by either swelling, filling in void space of the polymer with analyte, or dissolving into the polymer to form a binary mixture. The interferometric response due to swelling dominates when the film is thinner than the evanescent field, that is, less than approximately 5,000 Å. Swelling produces the largest possible signal when the film's thickness is equal to the $1/e$ thickness of the evanescent field, approximately 1,100 Å. Figure 15 shows the relative signal caused from a uniform swelling displacing either air or water in a sensing application. The change in n_{eff} is calculated for a 10% increase in polymer thickness. The waveguide used in this calculation is SiO_xN_y (1,500Å, $n = 1.85$ on fused silica). Since these films are very thin, the response due to swelling is very fast (subsecond in many cases) and, with no reaction occurring in the sensing film, the swelling reverses once the stimulus is removed.

Analyte adsorption swells the polymer film vertically, which incorporates more of the evanescent field as it swells. The interferometer response is the result of a positive change in the refractive index of the film, and a negative change in phase as the propagating light beam is slowed down by the expanding film. Swelling of a

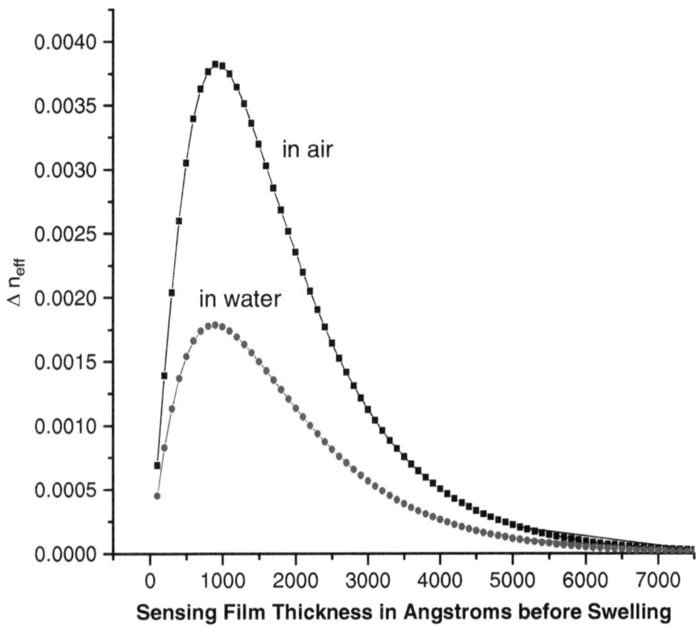

Fig. 15 Response due to 10% swelling in either air or water

polymer in air produces a larger signal than in water due to the relative differences in refractive indexes of air ($n = 1.0$) and water ($n = 1.333$) that are being displaced by the polymer's swelling.

Almost all polymers show a negative phase change when an organic solvent is present. However, a few solvent/polymer systems actually contract, thereby producing a positive phase change. Contraction of a film happens when a nonsolvent vapor interacts with the polymer film, a nonsolvent being a solvent that does not dissolve the polymer, has no affinity for the polymer, and in fact, makes the polymer retreat from the solvent front. Swelling is fast and produces large sensor responses; however, it is not a desirable mechanism for most sensing purposes owing to the fact that a portion of the evanescent field is exposed to the environment. This makes the sensor also sensitive to environmental background changes, such as bulk-index changes, scattering, or adsorption from particulate or colored matter or bubbles (in liquid applications), which may interfere with the measurement of the analyte of interest.

Fast response is possible with swelling-based sensing mechanisms. For example, the mapping of a chemical plume caused by a release into the air could employ such a sensor positioned on an unmanned aerial vehicle (UAV). With this vehicle traveling at 40 miles/h, subsecond response times would be required to locate and map out the released analyte. An example of this is shown in Fig. 16, which describes a thin film of poly(2,2-bistrifluoromethyl-4,5-difluoro-1,3-dioxole-*co*-tetrafluoroethylene), Teflon AF® placed on an interferometer; which in turn was placed in the nose cone of an UAV, which was positioned in a wind tunnel with the wind moving at 40 mph. Small amounts of toluene vapor were introduced into the air stream to get an idea of whether the sensor would work in such an application. The sensor responded quickly, established equilibrium in seconds, and reversibly returned to baseline after the material passed.

Two other passive mechanisms that have been studied which involve films that are thicker than the evanescent field: free-volume filing and dissolution. These thicker films shield the evanescent field from many environmental effects and offer greater sensing volume for increased sensitivity and selectivity.

Filling of the free volume occurs in polymers that have glass-transition temperatures that are higher than ambient temperature. It is thought that when these

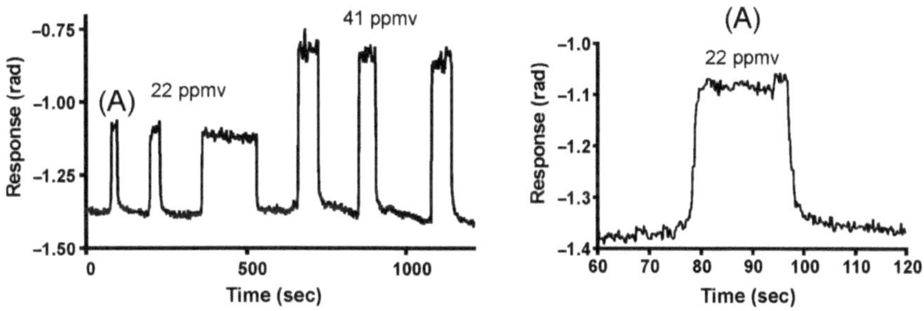

Fig. 16 Free-volume-filling response of Teflon AF® to toluene at 40 mph, detail of labelled response A (right)

polymers are cast on the surface of the sensor, interstitial voids are produced between the rigid polymer chains in the polymer into which small organic molecules can penetrate and fill. If the polymer and the analyte are not soluble in each other, this mechanism will dominate the response that causes a change in the refractive index of the film and produces an interferometer-sensor response. The filling will stop once all the available void space is filled, if the polymer has no solubility in the solvent doing the filling. In the void-filling mechanism, it is thought that the voids contain either air or water vapor, the analyte penetrates the pores and displaces the air or water. Since air and water have very low refractive indices, the response from this mechanism produces a positive change in refractive index, a negative change in phase. The response from a series of aromatic and chlorinated solvents in water being adsorbed into a Teflon AF® film in which only the free-volume-filling mechanism is permitted is shown in Fig. 17 [22].

The free-volume-filling responses are linear and all produce a positive change in refractive index or a negative change in phase. The analytes diffuse into the void volume in the Teflon AF® and do so reversibly, as shown in the time plot in Fig. 18, where a step function in concentration of various aromatics is introduced to the Teflon AF®-coated interferometer. Benzene, toluene, and xylene diffuse into the Teflon AF® from a solution containing known concentrations of the aromatic

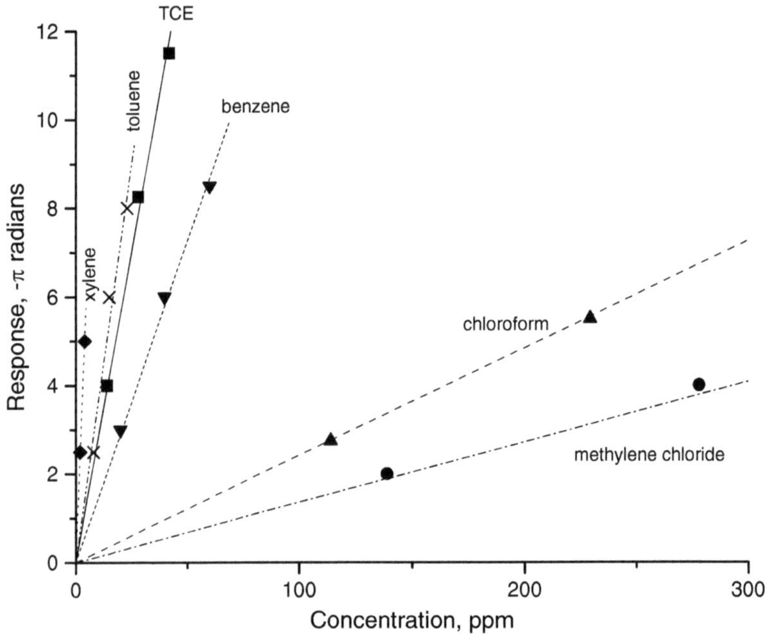

Fig. 17 Response curve to aromatic and chlorinated hydrocarbons in water using Teflon AF® sensing film. *Note*: the plots are shown in $-\pi$ radian on the y-axis corresponding to a positive change in refractive index

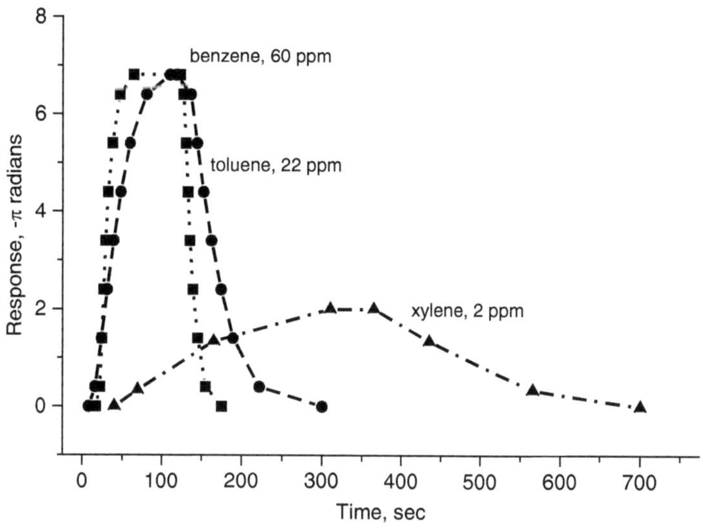

Fig. 18 Time response to a step change of concentration of aromatic hydrocarbons in water using Teflon AF® sensing film

hydrocarbons. Once equilibrium is obtained, the solution flowing over the sensing film is changed to pure water which extracts the analytes from the film in an manner analogous to a continuous extraction process. The shape of the sensor response as the aromatic hydrocarbons are diffusing into the polymer is driven by the partition coefficient and the flow rate. Flow is kept constant so that the response is solely a result of concentration and partition coefficient. The response due to refractive index differences are almost automatically normalized in this plot since the refractive index difference between the three compounds is only 0.004. If one plots the number of extractions required to reach equilibrium by the film, one finds that the lower the partition coefficient, the faster equilibrium is attained, i.e., the smaller number of extractions. Extraction out of the polymer almost mirrors extraction into the polymer.

The number of extractions required to extract xylene from the film is much greater than for extracting benzene out of the film, since the octanol/water partition coefficients differ by 1, a factor of 10, 2.1 and 3.16 for benzene and xylene, respectively.

There is a limited amount of void space available in a polymer for the adsorption of an insoluble small molecular analyte. If one exposes a solution containing an analyte small enough to be adsorbed by the polymer, and the polymer has an affinity for this analyte, the analyte will fill in these void spaces until the space is full. The amount that can be adsorbed can be determined interferometrically by increasing the concentration of the analyte containing solution. The signal from the interferometer ceases to increase at a concentration indicative of the amount of void volume in the polymer and the partition coefficient between the solution and the polymer film deposited on the sensing arm of the interferometer. This void space has been measured for Teflon AF® with two molecules of different molecular sizes and shapes: chloroform and toluene. With a 5,000 Å thick Teflon AF® film, the concentration of toluene and chloroform in a water solution were separately ramped up to saturation. The total phase shift resulting with toluene was -131π radian, and -140π radian for chloroform.

These responses correspond to a void volume of 9.8% using toluene as the void-filling compound in the calculation and 13.3% when chloroform, a smaller compound, is used in the calculation. The voids apparently have not only maximum capacity but also pores with some size-discrimination ability. These amounts correspond to 86% of the saturation limit of toluene in water and only 65% for that of chloroform. Larger molecules with even greater partitions were found not capable of being adsorbed into Teflon AF®. The assumption here is that the pore size in Teflon AF® prevents large molecules, such as naphthalene and bis(2-ethylhexyl phthalate), a plasticizer, from getting in.

Teflon AF® has remarkable solvent properties and acts like a solid organic matrix. However, most other polymers are going to have some interactive behavior between the analyte of interest (such as aromatic and chlorinated compounds) and the sensing polymer. Not only will an analyte diffuse into the pores of a high Tg film but will also adsorb into the polymer itself, thereby creating an interpore swelling effect and ultimately dissolving into the polymer itself. The relative refractive index change of these various mechanisms, although somewhat predictable, can lead to complicated responses which can limit the utility of this mechanism except at low concentrations of the analyte.

An example of such a biphasic response was observed using the copolymer poly (isobutylmethacrylate–*co*-*n*-butylmethacrylate) as a sensing film. The film has a refractive index of 1.481 and a *T*g above room temperature. Chloroform in water was slowly ramped up into a solution flowing over the methacrylate film. At low concentration, only the free-volume filling, with a positive refractive index change is seen. As the concentration of chloroform is increased, the chloroform begins to be adsorbed into the polymer's internal surfaces, swelling the pores shut and is producing a positive refractive index change. However, as more chloroform added and is adsorbed by the polymer, the polymer/chloroform film begins to form a binary mixture whose refractive index is being determined by the relative concentrations of methacrylate and chloroform ($n = 1.448$). As still more chloroform is added, the binary mixture's refractive index begins to lower, passing through zero on the way to an equilibrium value much below the starting point. Switching back to pure water extracts the chloroform out of the polymer film, but the sensor simply returns to zero without going through the positive-indexed void-filling pathway. This hysteresis effect would make certain analyte/polymer pairs difficult to characterize. If an analyte with a refractive index higher than the polymer is being sensed, the different mechanisms are additive but not uniformally so. By selection of the correct combination of polymer/analyte, this set of mechanisms can provide a sensor that responds very fast at low concentrations, since the analyte can move quite freely into the deepest reaches of the polymer where the evanescent field is most concentrated. As the concentration increases, the response shifts to the other mechanisms, providing a dual response sensor.

Teflon AF® was used to measure freons, trifluoroiodomethane and trifluorobromomethane in buffered solution made to mimic blood serum (where the use of this sensor was intended) [33]. Interferometric responses proved to be much larger than expected when one considers the refractive indexes of the freons and the sensing films. Since the freon is being adsorbed into the free volume of the polymer and is displacing either air or water vapor, the resulting refractive index addition can account for the very large signal. The signal was 48 radian for 3.0 ppm and 135 radian for 14.2 ppm of trifluoroiodomethane in HEPES buffer and was totally reversible. The response was linear from the 3 to 20 ppm levels that were measured. The signal is apparently not due to displacement of liquid water; if it were displacing liquid water, the signal would have been much less due to the relative similar values of their refractive indices, which are 1.357 and 1.379 for trifluorobromomethane and trifluoroiodomethane, respectively.

Dissolution becomes the mechanism producing additional phase shift after all the voids of a high-Tg polymer are filled and the concentration of the analyte in the sensing solution still continues to increase. Without voids to fill, as would be the case with a low-Tg polymer, the dissolution mechanism becomes the first and only mechanism for a film that is thicker than the evanescent field. The dissolution mechanism provides a sensing film with the best discriminating ability of any passive mechanism.

Dissolution is a term used to describe the adsorption of an analyte by a polymer which has nowhere to go but to dissolve into it. Low-glass-transition-temperature polymers have no free volume, so an analyte that the polymer has an affinity for will be adsorbed into the bulk of the polymer, and then diffused throughout the polymer material until an equilibrium is reached. Since there are no pores or paths to allow the analyte to quickly pass through the polymer matrix, the material moves more slowly though the polymer than what is observed in a high-Tg polymer. Polymer chains have to move in order to allow the analyte molecules to pass through the sensing film. However, since the mechanism involves the simple mixing of materials with different refractive indexes, the interferometric result can easily be predicted and the polymer designed to produce either a positive change in refractive index as the analyte is adsorbed (if the analyte's refractive index is higher than the polymer's), or a negative refractive index change (when the analyte's index is lower than the polymer's), or even a null response (when the refractive indexes of the analyte and polymer are matched). The magnitude of the sensing film's response is a product of both the refractive index difference between the analyte and the polymer and the partition coefficient between the polymer and the analyte which determines how much of the analyte will be adsorbed by the polymer at equilibrium.

The dissolution mechanism can provide a strong discriminator of the analytes being detected in a sensing medium. Polymers can be chosen which have very similar partition coefficients for a given analyte and thereby using the refractive index difference will be the means for determining the identity of the analyte present in the media. For example, Fig. 19 shows the relative responses of three

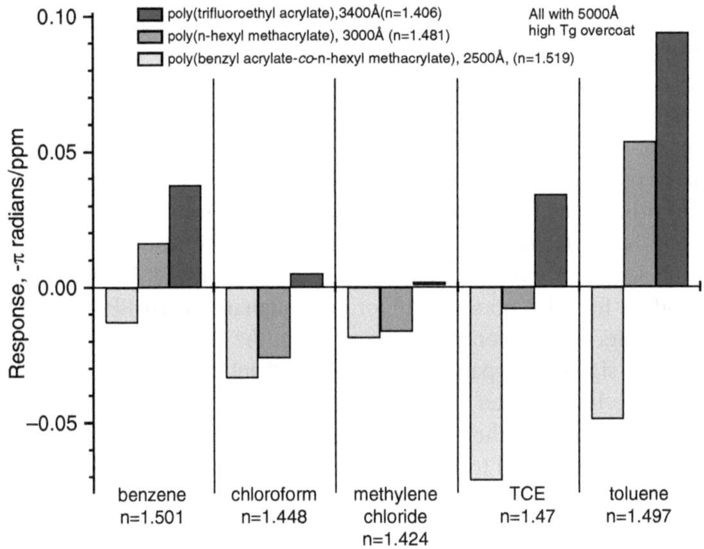

Fig. 19 Patterned responses from a polymer array using the dissolution mechanism

polymers to aromatic and chlorinated solvents in water. The polymers, a series of acrylate/methacrylates, have comparable partition coefficients for each individual analyte but differ in refractive index, which leads to the varied responses. The lowest indexed polymer, poly(trifluorethyl acrylate), has a refractive index of 1.406, lower than all the analytes measured, benzene (n = 1.501), chloroform (n = 1.448), methylene chloride (n = 1.424), trichloroethylene (n = 1.47), and toluene (n = 1.497), leading to a positive change in refractive index in the polymer film when the analyte is adsorbed into it. For the high refractive index polymer, poly (benzyl acrylate-*co*-*n*-hexyl methacrylate) (n = 1.519), the refractive index is higher than of all the analytes used, producing a negative refractive index change in the film when the analyte dissolves in the polymer. However, poly(n-hexyl methacylate) (n = 1.481) has a refractive index that lies between the refractive indexes of aromatic and chlorinated solvents, thereby producing (i) a composite positive refractive index change and a negative phase change when the aromatic hydrocarbons are being adsorbed into the polymer film and (ii) a composite negative index change and positive phase change when the chlorinated solvents are adsorbed into the sensing film.

The measurement of the response to a chemical compound with a single polymer film can result in either a positive or a negative refractive index change, but it cannot be determined whether the magnitude of the response was the result of a small amount of material mixing into the polymer in which a large refractive index difference between the film and the analyte exists, or a large amount of analyte dissolving into a film where the refractive index difference between analyte and film is small. With two interferometers with sensing films of different refractive indexes and comparable partition coefficients, the ratio of responses of the two interferometers can determine the refractive index of the analyte present. A table would allow the identification of the unknown contaminant to be identified. As additional analytes are added to the mixture and partition into the array of polymers used, a straightforward mathematical analysis can determine the composition of the analytes present. The interferometric response should be additive, but the number of polymers used has to be one more than the number of analytes present. At low concentrations, responses appear to be additive so that, by knowing the refractive index and the partition coefficient of the analyte for a given polymer, the composition of mixture can be determined. It has been noted that, with the series of polymers tested, the partition coefficients do track to about 10% with the published octanol/water partition coefficients. With predictable partition coefficients, the refractive index differences of the polymer and analyte become the strongest parameter for determination of an analyte's identity. As the mixture becomes more complex, a more chemometic approach may be necessary to determine what is present if a mixture is encountered and especially if the responses deviate from being simply additive.

Teflon AF® or other materials which dissolve in a limited number of solvents could serve as good protection for a sensing film that may encounter a pure analyte in its measurement use. A pure solvent could dissolve away the sensing film. Cross-linking the sensing polymer could limit the polymer's solubility in pure solvent, but cross-linking has been shown to restrict adsorption and, therefore, the sensing capability of the film. Alternatively, the placement of a film such as Teflon AF® film or a sol-gel film could protect the sensing film while allowing the analyte to pass through to the sensing film. This theory was validated with a sensor coated with a laminate of a methacrylate film followed by a Teflon AF® coating. The sensor responded to an analytical solution of benzene in water and was then placed in pure benzene overnight. The sensor was removed from the benzene and allowed to dry out. When the sensor was placed back in solution and exposed to the same sample of benzene in water as before, it produced the same response as measured prior to the exposure to pure benzene. Materials such as Teflon AF® or sol-gels which can also be fluorinated present good coating materials for use in sensing environments where pure solvent or dirty conditions may be encountered

A scheme of how different polyacylate/methacrylate films can be used to differentiate the identity of various chlorinated and aromatic hydrocarbons by using the dissolution mechanism alone is shown in Fig. 19. However, similar information can be obtained by employing a laminate of polymers that have contributions from both the dissolution and free-volume-filling mechanism. The results from such a system are shown in Fig. 20, in which the underlying film, poly(*n*-hexyl methacrylate), undergoes the dissolution mechanism, while the overlaying film, Teflon AF®, makes contributions to the overall signal through the free-volume-filling mechanism. The sensor's overall response is a combination of the $\pm$ refractive index change of the poly(*n*-hexyl methacylate) film combined with the always positive change of free-volume filling in the Teflon AF®. It should be noted that, in the previous example, the contribution of the Teflon AF® was minimal, if any, since this film resides almost entirely outside the evanescent

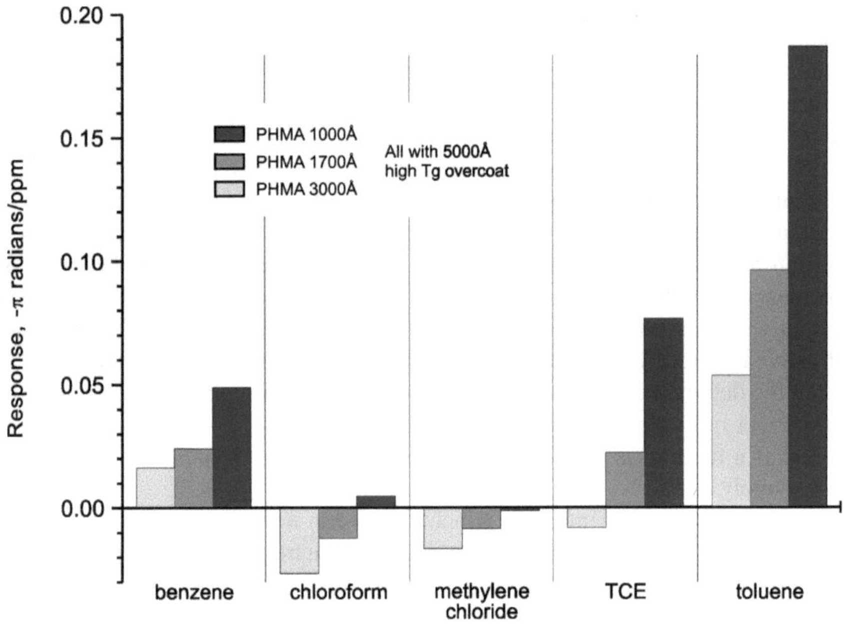

Fig. 20 Patterned response from combined free-volume-filling and dissolution mechanisms

field; whereas, in this example, the Teflon AF® film plays a major role in the response of the sensing film. Adjusting the thickness of the methacylate film alters the contributions from the two different mechanisms.

The dissolution mechanism appears on first examination to be the only mechanism capable of providing a ratio of responses from two different film/interferometer combinations that could lead to the determination of the refractive index and, therefore, the identity of the analyte present. This turns out not to be the case. If one looks at another set of data obtained from two related polymer-sensing systems that employ the free-volume-filling mechanism, one film, a laminate containing a underlayer of 400 Å poly (2,6-dimethyl phenylene-p-oxide), with Teflon AF® covering this polymer to a thickness slightly greater than 5,000 Å, and a second film comprised of only Teflon AF®, one can distinguish similar refractive index differences due to the ratio of response to trichloroethylene and its breakdown products, *cis*- and *trans*-dichloroethylene and vinyl chloride [34]. The response of these two films is shown in Fig. 21a and the ratio of the responses in Fig. 21b. What is interesting to point out is that the two films can distinguish between geometric isomers, *cis*- and *trans*-dichloroethylene, which do have different refractive indexes and different partition coefficients.

One may have noticed that a thin layer, just 400 Å thick, of poly (2,6-dimethyl phenylene-p-oxide) under Teflon AF® added greatly to the response of the film from 1.3 radians/ppm for trichloroethylene to almost 13 radian/ppm. The poly(2,6-

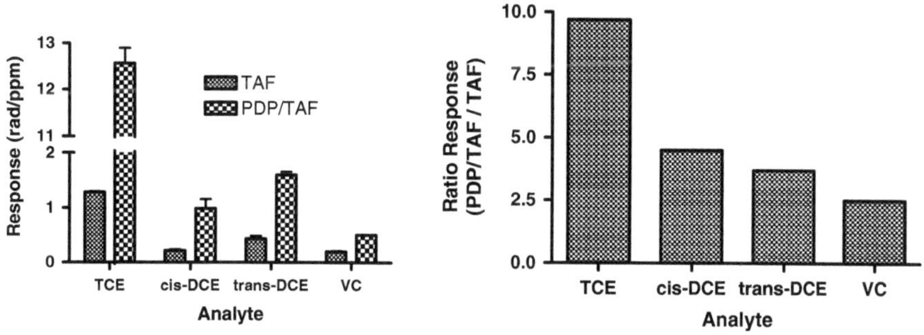

Fig. 21 Response and ratio of responses of two polymer-sensing systems for TCE and daughter compound measurements

dimethyl phenylene-p-oxide) film alone produce 250 radian/ppm, a response that is believed to be all passive in nature but that makes one wonder if something more is taking place in this material. Production of this magnitude of response would require a concentration of greater than 100,000 ppm adsorbed into the film for a partition coefficient of 5.1, which is not unheard of but high for TCE. This polymer also seems to have high affinities to the other chlorinated ethylenes, tetrachloroethylene, *cis*- and *trans*-dichloroethylene, and vinyl chloride, but very low affinity for chloroform and methylene chloride, which again makes one wonder if something other than mass addition is occurring with these chlorinated ethylenes and this polymer.

7.2 Active Sensing

Countless number of reactions could find use in active chemical sensing of a wide variety of chemicals. Examples include acid/base reactions, π–π complexation and covalent bond formation, among others, depending on the reactive nature of the targeted compound. These reactions have been shown to produce large interferometric signals through the formation of a different chemical species within the sensing film that has a substantially different refractive index from those of the starting materials. To date, only a limited number of active sensing mechanisms have been explored. Biosensing would be classified under the category of active sensing, including antibody–antigen interactions for the detection of proteins, toxins, viruses, and whole organisms; DNA hybridization and aptamer binding to organic biomolecules have been reviewed previously [6].

A challenge in active sensing is the design of reversible sensing chemistries. The available literature is full of chemical reactions that lead to an end-product. There are fewer examples of reactions known to be reversible. Reversibility may not be an issue if the compound one wishes to sense is not likely to be encountered, and if it was detected, evacuation of the vicinity would be required—such as in sensing for a chemical nerve agent. In this case, the contaminated sensor could either be held as a piece of legal evidence or be decontaminated and/or destroyed. However, for continuous monitoring of a fluctuating concentration of a targeted chemical such as in a flowing stream, mapping with a robotic sensor, or monitoring air intake to a building, a reversible real-time sensing film would be needed.

Glass waveguides have hydroxyls present at the waveguide surface that can respond to acid/base changes in the medium over the waveguide. To test this hypothesis an ion exchange waveguide made from BK-7 glass was fitted with a flow cell designed to hold a miniature pH electrode. This flow cell was placed over the sensing arm of an interferometer and a 0.01M solution of phosphoric acid was flowed over the waveguide. A 0.01M solution of sodium hydrogen phosphate solution was used to titrate the phosphoric solution reservoir supplying the solution to the waveguide. The pH electrode was positioned to measure the pH of the solution directly above the waveguide. The resulting titration curve shows the pH response due to the deprotonation of the hydroxyl protons on the waveguide surface. The response was rather small, 0.53 radian, but one would not expect much from an ion-exchange waveguide with an inherent sensitivity of 0.7 radian for a 0.001 change in refractive index over 1 cm pathlength. In fact, this is a rather large change when one realizes it is due to the deprotonation of only a monolayer of hydroxyls. It is also interesting that the pKa of the glass surface would be approximately 7.0, lower than the pKa_1 for salicic acid, but then the hydroxyls are, no doubt, modified by the silver ions concentrated at the waveguide surface.

Another set of experiments shows that the use of these surface hydroxyls is directly involved in producing a change in refractive index due to their deprotonation by base. Another ion-exchange waveguide produced from BGG-31 glass in which the sodium ions were exchanged for cesium ions in a molten-salt bath showed a reversible reaction with ammonia in air. Ammonia is rapidly and reversibly deprotonated in air by the available hydroxyls on the waveguide surface. The response to ammonia saturated at between 1 and 2 parts-per-million and produced a 0.62-radian phase change over 1.5 cm pathlength. The total response of the ammonia and the pH response produced fairly comparable phase shifts, when all the hydroxyls were deprotonated on the waveguide surface.

Two experiments show the protonation of ammonia to be the mechanism occurring to produce the phase response with exposure to ammonia. One experiment involved simply washing the surface with a strong base, potassium hydroxide, which converted all the hydroxyls to potassium alkoxides. Exposure of this surface to ammonia showed no response due to the lack of free hydroxyls remaining after base wash. In the other experiment, the free hydroxyls were converted to trimethylsilyl groups through the covalent bond-forming reaction with hexamethyldisilazane (HMDS). Again, exposure to ammonia produced no response.

Early work using active chemistry centered on acid/base-type sensing chemistries for either the monitoring of pH or the sensing of ammonia or other amine-type bases. These initial results led to the development of an optical pH sensor illustrating some of the capability the interferometer has of responding to reactive changes occurring on the waveguide surface [36]. A silicon nitride waveguide was

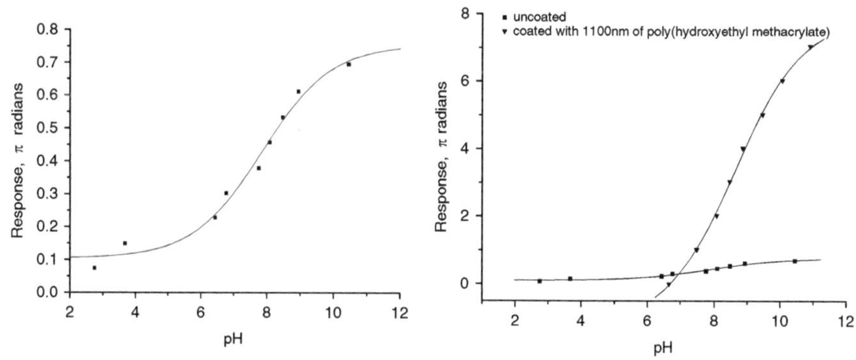

Fig. 22 pH responses of a monolayer of glutamic acid without (left) and with an overlaying proton conducting film (right)

derivatized with a covalently attached coating of the amino acid glutamic acid. A flow cell was positioned on the waveguide along with a pH electrode to monitor the pH titration of the solution passing over the waveguide and the reference arm was rendered inert with HMDS. With increasing pH, the glutamic acid groups on the waveguide became deprotonated producing an interferometric response. The response was small, 0.8π radian over a 6–10 pH range, using a waveguide with 4–5π/0.001/cm sensitivity, as shown in Fig. 22a. Interestingly, the p*K*a of the glutamic acid was shifted from the acid side to slightly basic. The issue remains "Is the response being measured due to the carboxylic acid being shifted up in its p*K*a?" or "Is the amino group being shifted down since the amino group is what is producing the response?" Even more interesting were the results from the experiment with the placement of a proton-conducting polymer over the glutamic acid. The polymer used was poly(2-hydroxyethyl methacrylate), a polymer that contains a good deal of water but is not soluble in water. Over the same pH range, a response was recorded with a tenfold greater signal than without the polymer, and the polymer coating showed no time delay in response; see Fig. 22b. It has been conjectured that the polymer allows for a proton cascade to occur through the film when the pH outside the film changes and is capable of pulling a proton from of the monolayer film by proton transfer through the film. The polymer also prevents the counterion from diffusing through so it remains outside, thereby increasing charge separation. It is hypothesized that this charge separation couples with the evanescent field to produce an enhanced response.

Other factors could be the cause of this enhancement, such as water movement through the film solvating the resulting ions, but it is conjectured that this would be a relatively slow process. For now, the underlying reason for the sensing enhancement remains unclear. However, other data support the current working hypothesis.

For example, placing a base-sensitive polymer, poly(4-vinylphenol), under the water-containing poly(2-hydroxyethyl methacrylate) produces a response for pH in a water solution over the laminate. However, if the water containing polymer is exchanged for a polymer that does not contain water such as poly(methyl methacrylate), the pH response is eliminated. The response is recovered by utilizing ammonia to adjust the pH. The theory here is that ammonia is able to diffuse through the polymer without the need for water and reacts with the acidic phenol. However, with alkaline hydroxide, base diffusion is not possible and the mechanism shuts down. Questions remain as to the nature of these acid/base reactions.

These acid/base reactions are also among of the most obvious reactions that can be used for chemical sensing. The sensing film can be basic in structure and reacts with an acid in the sensing medium or an acid-containing film can reacts with a base in the environment. The first example of one of these reactions is also one of the first optical-waveguide-sensing experiments reported. In this experiment, an optical fiber was coated with an oxazine dye [37]. When this fiber is exposed to ammonia, the ammonia reacts with the dye to produce a color change which in turn alters the light throughput through the fiber. This reaction is analogous to the interferometric measurement but whereas a color change alters light throughtput in the fiber, in the interferometer, the same reaction can cause a change in the refractive index, which is measured as a phase change. If the oxazine dye is used with the sensing film for detection of ammonia in an interferometric configuration, a response due to the refractive index change in the oxazine can be detected. Interferometric measurements require no color change such as that produced with the oxazine/ammonia reaction, which allows for a much greater variety of reactions to be used with the interferometer to sense a chemical analyte.

A better understood acid/base sensing reaction can be found in the detection and monitoring of ammonia either in air or in water. Two systems have been explored in great detail: one using citric acid as the sensing moiety and the other using poly (4-vinyl phenol).

The citric-acid case is one that points out the amplification possible from a chemical reaction—in this case, proton transfer—and the ability to design a reference to cancel out much interference, thereby permitting the interferometer to zero in on the measurement of one analyte. Ammonia is a weak base; the p*K*a of the ammonium ion is 9.8, and it will react reversibly with a weak acid. Citric acid is a triprotic acid, with p*K*a's = 3.1, 4.8, and 6.4. The least acidic proton on the citric acid would be a good candidate for a reversible reaction with ammonia. The other two may be too acidic and may not reverse after exposure.

To provide a sensing matrix that can use the least acidic proton on citric acid, polyethyleneimine, with p*K*a = 11, serves as the matrix onto which citric acid will be ionically tethered to the polymer's backbone. Titration of a solution of polyethyleneimine with a solution of citric acid to a pH of 6.0 produces a polymer solution of polyetheneimine-citric acid, where the lone least-acidic proton is on the citric acid available for reaction with the ammonia, while the other two are bound to the amines of the backbone of the polyethyleneimine. For a reference arm, another solution of polyethyleneimine is titrated with a solution of citric acid to a pH of 8, thereby yielding a solution in which all the protons of the citric acid are on the polyethyleneimine backbone. The first solution, pH 6.0, is deposited on the sensing arm of a silicon nitride waveguide. The second solution, pH 8.0, is deposited on the reference

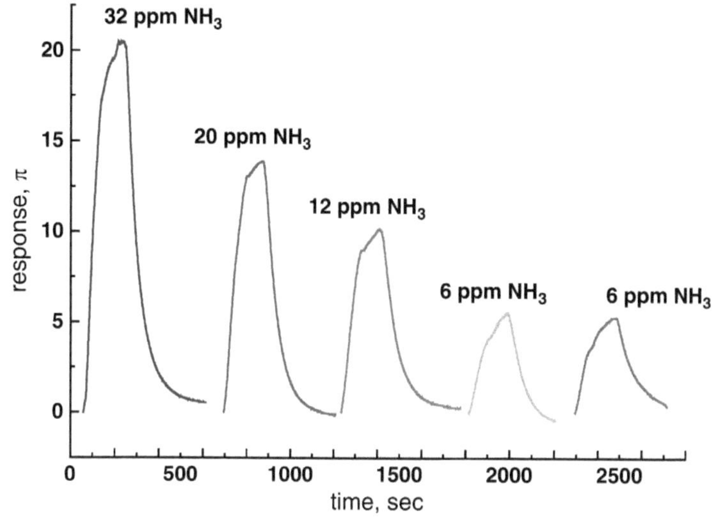

Fig. 23 Interferometric response to various concentrations of ammonia in air

arm. When the sensor is exposed to ammonia, ammonia will diffuse into both arms of the interferometer equally since they differ very little in chemistry; in fact, there are only a couple of drops of citric-acid-solution difference between the two solutions. The mass addition to both arms is equal and would cancel out optically if not for the citric-acid protons available for exchange with the ammonia in the sensing arm.

The sensor's response resulting from the protonation of the ammonia within the sensing film, approximately 1 π radian/ppmv of ammonia, using a waveguide with 4–5π/0.001/cm sensitivity, is shown in Fig. 23. The response is also reversible once the ammonia stimulus is removed, the citric acid regains its proton and the ammonia diffuses out. This sensor exemplifies the effect that a dipolar change in the sensing field can produce, which far outdoes a simple mass change. The mass addition of ammonia to the pH 8.0 film was measured and found to be less than 1 π radian for 500 ppmv.

The advantage of having a reference arm that is almost identical in chemical composition allows for the cancellation of other interfering effects. The effect of humidity has no bearing on the interferometer unless the humidity drops below 20% and the reaction stopped (making it apparent that water is necessary). In this case, the ammonium ion and the citrate ion may also be appreciably separated, adding to the dipole effect. As an added bonus, the similarity between the two arms also eliminates any solvent effects since they partitioned equally in the two arms at the levels measured.

The other system studied for the measurement of ammonia is one using poly (4-vinylphenol) as the acidic-sensing group. Phenol is a weakly acidic compound, pKa = 9.9, that will react reversibly with ammonia. Sensing films comprised of poly(4-vinylphenol) produce about 1 radian of phase change per 3 ppmv of ammonia using a waveguide with 4–5π/0.001/cm sensitivity. The advantage of the poly

(4-vinylphenol) is that it can be used in aqueous-solution measurements if ammonia concentrations are not very high. Otherwise, the film needs to be placed under a water-insoluble polymer to prevent loss at higher concentrations of ammonia due to the solubility of the poly(4-vinylphenoxide) product. The solid poly(4-vinylphenol) polymer is much more durable than the viscous citric acid-polyethyleneimine polymer films unless the citric-acid-based material is placed under a protective coating. The poly(4-vinylphenol)-based sensor shows little response to organic solvents or water and can be referenced with either poly(4-vinylphenoxide) or poly(4-vinylanisole), among others.

The poly(4-vinylphenol) is not of ideal acidic strength for ammonia sensing but finds better use for alkyl amine (an order of magnitude stronger base than ammonia) sensing. Poly (4-vinylphenol) works well in detecting methamphetamine, a compound of interest in law enforcement; however, methamphetamine is typically encountered as the acid salt. To use this polymer sensing film for the acid salt of methamphetamine, it was found that the solid acid salt powder can be collected and passed through a strongly basic-ion-exchange resin that converts the acid salt to the free amine for subsequent detection with a poly(4-vinylphenol)-based sensor. Detection levels in the parts-per-billion have been demonstrated using silicon nitride-based, 3-5π/0.001/cm waveguides.

Another preconditioning scheme has been used to measure chlorine in surface water followed by another active sensing scheme based on π-complexation. Chlorine does not as easily form π-complexes as iodine does. On passing a solution to be assayed for chlorine concentration through an ion-exchange resin that is in the iodide form, the chlorine converts the iodide to iodine that then gets carried along in solution and diffuses into and complexes with polystyrene sensing films. This reaction is reversible but the resin will be used up in time and require replacement, but that is a small inconvenience. It is also possible to have the iodide ions in the sensing film, leaving the chlorine to diffuse into the sensing film, convert the iodide to iodine, and to have the iodine complex with the aromatic styrene. However, the sensing film's available iodide ions would be used up over time and the sensing chip require changing, as opposed to a simple cartridge at the sensor input. Sensing capabilities for chlorine are in the 10's to 100's parts-per-billion range. Note that measuring chlorine in aqueous solution is not a simple measurement of the chlorine species as one has to account for other species present such HOCl and the hypochlorite ion, depending on the pH of the solution. Sensors designed for each of the possible species are needed for the total analysis of chlorine and related materials in solution.

Acid/base chemistry is also an obvious reaction path for selective detection of a variety of materials. For example, acid/base reaction can be made to occur within a molecularly imprinted pocket to assist in differentiating the molecule of choice before deprotonating it to produce an ion pair that then interacts with the evanescent field. This approach is currently being used to selectively detect TNT to levels in the low parts-per-trillion. 2,4,6-trinitrotoluene, a potent explosive, is also a weak acid having a p*K*a of approximately 14.5 [38]. A TNT derivative is synthesized with a tether to a silane. The silane is chosen so the group can be

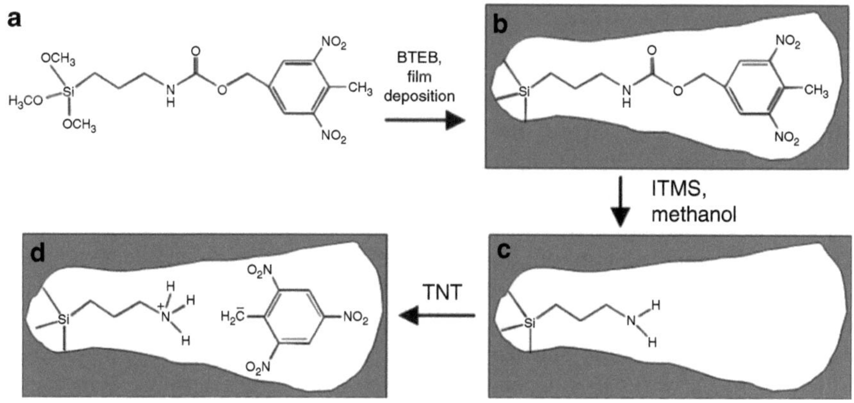

Fig. 24 Molecular imprinting strategy for the detection of TNT (see text for details)

incorporated in a sol-gel condensation reaction which incorporates the TNT-shaped molecule in the three-dimensional matrix of the sol-gel. After the sol-gel is deposited on the waveguide and cured, the TNT moiety is selectively removed leaving behind a pocket in the sol-gel topologically defined to fit and molecularly interact with a TNT molecule. Present in this molecular pocket is an amine group capable of abstracting the proton from TNT to form the anion. The sequence is illustrated in Fig. 24 [35]. The mass addition and, to a greater extent, the proton abstraction produces a dipolar change in the film that interacts strongly with the evanescent field of the interferometer. Responses of 1 radian per 100 parts-per-trillion for a $4\pi/0.001$/cm waveguide have been measured and shown to be reversible.

These few examples of acid/base and π-complexation reactions illustrate the responses and the low detection limits possible when an active-sensing chemistry is employed in an interferometric sensor. Many reactions are available to select compounds that one may want to detect. A few of the reversible ones have been shown as reversibility may be desired when one is trying to measure the real-time changes in a changing environment. But for the detection of a potentially deadly hazard such as a chemical warfare agent, an explosive, or some toxic industrial compounds, reversibility may not be as imperative as detection. Also, an irreversible reaction allows the integration of a compound that is in very low concentrations up to a level where detection is possible. Of the reactions known to most organic chemists, practically all are irreversible since that is what chemists report. There is no journal of reversible chemical reactions. With the large library of reactions available to chemists, any number of these can be incorporated onto a waveguide chemical sensor and set to look for and assay any number of compounds, where having the reaction reversible may not be essential. If the waveguide can be regenerated after detection or if the waveguide can be quickly exchanged, an irreversible reaction may be more beneficial.

7.3 Thermal Response

Interferometers have long been used for better or worse as thermometers. Applying different materials on each arm of a Mach-Zehnder or Young interferometer yields a device acutely sensitive to temperature. Since every material undergoes a change in refractive index with a change in temperature, it is difficult to remove the thermal response of an interferometer from the response produced by placing a chemically selective film on one arm of the interferometer. Solutions include taking the measurement while holding the temperature constant, designing the sensing and reference arms to experience the same $\mathrm{d}n/\mathrm{d}T$ (doable but difficult), or by measuring temperature independently and removing the temperature effect in the software. Which approach works best depends on the application. Interferometric measurements can usually be done as a relative measurement by switching between the sample medium and the sensing medium which has been filtered to remove the analyte being measured. This approach has worked well in the sampling of groundwater where the water being sampled was passed through an activated charcoal filter and was used to establish a baseline. Simply bypassing the filter brings the water to be measured to the sensor. It is easy to switch back and forth between the filtered water and sample water, and clean water is not needed at the site being measured.

For applications where wide temperature ranges occur, the interferometer can be calibrated for temperature response alone. A plot of the sensor's temperature response can be stored. By using a thermometer or temperature-measuring device the interferometer can be corrected for $\mathrm{d}n/\mathrm{d}T$ effects of the sensor. The most favorable approach is to develop a sensing film that has a very large response to the analyte one wants to detect, a response that greatly overrides any change caused by temperature during the measurements. However, if very accurate or very high sensitivity measurements are required, then the temperature needs to be independently measured and corrected and not just for $\mathrm{d}n/\mathrm{d}T$ effects in the sensing and reference arms of the interferometer but also for any changes in the response of the sensing film to a given analyte with temperature.

7.4 Massless Response of the Interferometer

The addition of mass provides the means of transduction for many chemical sensors, including surface acoustic wave (SAW) devices, quartz crystal microbalances (QCM), and microcantilevers. In all these devices, the mass addition either perturbs the vibration, oscillations, or deflection within the transducer. The mode of transduction in an optical interferometer can also be linked to mass addition; the sensor's response is altered by refractive index changes in the material being monitored. It is possible that this change can be elicited solely from refractive index changes without the addition of mass, although in sensing a particular compound, mass addition typically takes place. Related refractive index changes can amplify the transducer's signal, thereby making the sensor much more sensitive to the analyte one wishes to measure.

Two examples, polymer polymerization and double-band isomerization, will illustrate the inherent mode of measurement that can result without the addition of mass. An early example used to test the feasibility and sensitivity of a two-mode interferometer [26] measured the refractive index changes at the surface of a planar waveguide by monitoring the polymerization of a UV-curable epoxy. The monomeric epoxy was spread out on the multimode waveguide interferometer in which two modes were optically interacted. UV light was used to initiate and sustain the polymerization. As the epoxy cured, the refractive index of the epoxy increased and a large interferometric signal resulted. In this example, there was no mass change in the film since all the organic components became incorporated in the polymerized epoxy film. There may have been some shrinkage of the film but the major response was due to refractive index changes. It should be noted that the same experiment performed on one of the acoustic transducers would probably also produce a transducer response due to the stiffening of the material as it polymerized.

In the second example, the acoustic transducers probably would not produce a response since the change is localized: a geometric molecular change. This example involved no change in mass but did produce an interferometric response. A *cis*-azobenzene was admixed with a polymer and coated onto a Young interferometer in which one arm was coated with the azobenzene/polymer mix while the other arm was covered in silicon dioxide to provide an inert reference arm. When UV light was turned on, the *cis*-azobenzene was converted to *trans*-azobenzene. Since *cis*- and *trans*-azobenzene differ in refractive indexes, the refractive index changed proportionally to the amount of azobenzene present that was converted to its geometric isomer. When white light was allowed to shine on the film, the isomerization reversed and the interferometer signal returned to its origin. No mass addition occured; however, a refractive index change did occur, producing an interferometric response.

The azobenzene experiment was taken a little further. The same concentration of *cis*-azobenzene was added to two similar polymethacrylate films. These polymers differ only in their glass transition temperature with one having a Tg higher than room temperature and the other lower than room temperature. When the UV light was turned on, the higher-Tg mixture responded faster and produced a larger signal than the low-Tg mixture. The hypothesis being proposed is that azobenzene molecules residing in the void space of the high-Tg mixture are able to isomerize, that is, rotate to a greater degree than the low-Tg mixture in which the rotation encounters other atoms in the more compacted low-Tg mixture.

These examples show that mass addition is not essential for an interferometer's response. Such schemes could find use in sensing. For example, a monomeric sensing film could be used to amplify the response to an analyte capable of directly or indirectly leading to polymerization. A sensing film composed of a monomeric species could remain unperturbed until a free radical initiator was detected. A peroxide such as triacetone triperoxide (TATP), the explosive whose use was attempted by the "shoe bomber", could be detected by such a scheme in very small amounts. The peroxide could be adsorbed by the monomer containing sensing film that is also doped with a catalyst that would decompose the TATP to free radical constituents leading to polymerization of the sensing film. A small amount of peroxide could lead to a large resulting signal by incorporating such a sensing scheme.

8 Concluding Remarks

Optical interferometers hold the promise of chemical and biological sensors capable of detecting a wide variety of analytes. Chemistry can be passive or active in the manner it interacts with the analyte one wishes to detect. Passive chemistries have been well defined for most polymer-based adsorption-based mechanisms. Active chemistry offers an unending array of chemical reactions that can be tailored to react with a given analyte that one wishes to detect. Passive chemistry is employed when the analytes one needs to detect do not have any particular chemistry that one can take advantage of. But one can rely on the distinctive refractive index that each and every material possesses along with its distinctive partition coefficient that determines how much material will be adsorbed into a sensing film. Taking advantage of being able to tailor the refractive index of a sensing film, the glass transition of a given polymer material and the chemical features of the adsorbing film will enable some degree of discrimination and sensitivity based on these parameters. Using these different passive sensing mechanisms in concert will, through the use of several interferometers, allow a sensing array to provide some degree of identification and quantification of mixtures as long as the mixture does not become too complex in nature.

In many cases, one may only encounter one analyte at a time or one analyte in majority. These are the cases where a sensor can provide an inexpensive alternative to a complete chemical analysis which is costly, time-consuming, and cannot provide real-time analysis.

If the analyte of interest has some reactive functionality on the molecule, the application of interferometric sensing becomes easier through active sensing modalities. Not only does active sensing provide greater identifying capability but the refractive index change from a chemical reaction is usually much larger than that from a passive addition of two materials of different refractive indexes mixing.

Passive mechanisms rely on the concentrating ability of a film to adsorb analytes from the air or solution to produce rather large changes in refractive index. Combining this concentrating ability with some type of chemical reaction or interaction, interferometric responses can be produced which correspond to concentration in the low parts-per-billion to parts-per-trillion levels. These levels of sensitivity were never thought possible until the various chemical interactions described in this chapter were uncovered. With improved sensing chemistry, not only improvements in detection levels will be accomplished but also better means of identifying what is present. Ultimately, differentiation of mixtures of chemicals will be enabled with devices that merely rely on the interaction of an evanescent optical field with a chemical sensing film. Sensors based on this technology could find widespread application. Small, inexpensive, low-power, rugged sensors could find use in all sorts of applications, providing users with information to improve and possibly save lives simply through this measurement of the speed of light.

References

1. Young T (1804) The Bakerian lecture. Experiments and calculations relative to physical optics. Phil Trans R Soc Lond 94:1–16
2. Lukosz W, Tiefenthaler K (1983) Sensitivity of grating couplers as integrated-optical chemical sensors. In: 2nd European conference on integrated optics, Florence IEEE conference proceedings, vol 227, pp 152–153
3. Tiefenthaler K, Lukosz W (1984) Integrated optical switches and gas sensors. Opt Lett 9:137–139
4. Tiefenthaler K, Lukosz W (1985) Grating couplers as integrated optical humidity and gas sensors. Thin Solid Films 126:205–211
5. Campbell DP, McCloskey CJ (2002) Optical interferometric biosensors. In: Ligler FS, Rowe Taitt CA (eds). Optical Biosensors: Present and Future. Elsevier, Netherlands, pp 277–304
6. Campbell DP (2008) Interferometric biosensors. In: Zourob M, Elwary MS, Turner A (eds) Principles of bacterial detection: biosensors, recognition receptors and microsystems. Springer, New York, pp 169–208
7. Lambeck PV (2006) Integrated optical sensors for the chemical domain. Measurement Science and Technology 17:R93–R116
8. Kersey AD (1990) Recent progress in interferometric fiber sensor technology. In: Depaula RP, Udd E (eds). Fiber and Laser Sensors VIII Proc Soc Photo Opt Instrum Eng 1367:2–12
9. Nishihara H, Haruna M, Suhara T (1985) Optical integrated circuits, Chap 2, McGraw-Hill, New York
10. Gato L, Srivastava R (1996) Time-dependent surface-index change in ion-exchanged waveguides. Opt Commun 123:483–486
11. Millar CA, Hutchins RH (1978) Manufacturing tolerances for silver-sodium ion-exchange planar optical waveguides. J Phys D Appl Phys 11:1567–1576
12. Walker RG, Wilkinson CDW (1983) Integrated optical ring resonators made by silver ion-exchange in glass. Appl Opt 22:1029–1035
13. Nishihara H, Haruna M, Suhara T (1985) Optical integrated circuits. McGraw-Hill, US, p 226
14. Heuberger K, Lukosz W (1986) Embossing technique for fabricating surface relief gratings on hard oxide waveguides. Appl Opt 25:1499–1504
15. Ramos BL, Choquette SJ, Nell NF Jr (1996) Embossable grating couplers for planar waveguide. Opt Sens Anal Chem 68:1245–1249
16. Hartman NF (1990) Optical sensing apparatus and method. US Patent 4,940,328
17. Campbell DP (2005) Interferometric sensors for monitoring our environment. In: Proceedings of LAT conference, St Petersburg, Russia
18. Heideman RG, Kooyman RPH, Greve J (1994) Immunoactivity of adsorbed anti human chorionic gonadotropin studied with an optical waveguide interferometric sensor. Biosens Bioelectron 9:33–43
19. Heideman RG, Veldhuis GJ, Jager EWH, Lambeck PV (1996) Fabrication and packaging of integrated chemo-optical sensors. Sensors Actuators B: Chem 35:234–240

20. Brandenburg A (1997) Differential refractometery by an integrated optical Young interferometer. Sensors Actuators B: Chem 38:266–271
21. Brandenburg A, Krauter R, Kunzel M, Schulte H (2000) Interferometric sensor for detection of surface bound bioreactions. Appl Opt 39:6396–6405
22. Campbell DP, Moore JL, Cobb JM, Hartman NF, Schneider BH, Venugopal MG (1998) Optical system-on-a-chip for chemical and biochemical sensing: the chemistry. Proc SPIE 3540:153–161
23. Heideman RG (1993) Optical waveguide based evanescent field immunosensors. PhD dissertation, University of Twente, Netherlands
24. Stamm CH, Lukosz W (1993) Integrated optical difference interferometer as refractometer and chemical sensor. Sensors Actuators B: Chem 11:177–181
25. Stamm CH, Lukosz W (1994) Integrated optical difference interferometer as biochemical sensor. Sensors Actuators B: Chem 18:183–187
26. Hartman NF, Cobb JM, Edwards JG (1998) Optical system on-a-chip for chemical and biochemical sensing: the platform. Proc Soc Photo Opt Instrum Eng 3537:302–309
27. Koster T, Posthuma N, Lambeck P (2000) Fully integrated optical polarimeter. Europtrode V:179
28. Lillie JJ, Thomas MA, Denis KA, Jokerst NM, Henderson C, Ralph SE (2004) Modal pattern analysis and experimentatal investigation of multimode interferometric sensing: a path to fully integrated silicon-CMOS-based chem./bio sensors. Proc Lasers and Electro-Optics Society Annual Meeting –LEOS, pp 352–353
29. Heideman RG, Lambeck PV (1999) Remote opto-chemical sensing with extreme sensitivity: design, fabrication and performance of a pigtailed integrated optical phase-modulated Mach-Zehnder interferometric system. Sens Actuators B61:100–127
30. Hartman NF (1997) Integrated optic interferometric sensor. US Patent 5,623,561
31. Ymeti A, Kanger JS, Greve J, Lambeck PV, Wijn R, Heideman RG (2003) Development of a multichannel integrated Young interferometer. Appl Opt 42:5649–5660
32. Ymeti A, Kanger JS, Greve J, Besselink GAJ, Lambeck PV, Wijn R, Heideman RG (2005) Integration of micro fluidics with a four-channel integrated optical young interferometer immunosensor. Biosens Bioelectron 20:1417–1421
33. Jaio Z, DeJesus VR, Iravanian S, Campbell DP, Xu J, Vitali JA Banach K, Fahrenbach J, Dudley Jr SC (2005) A Possible Mechanism of Halocarbon-induced Cardiac Sensitization Arrhythmias. J Mol Cell Cardio 41:698–705
34. Campbell DP, Gottfried DS, Cobb-Sullivan JM (2004) Groundwater monitoring of VOCs with an interferometric optical waveguide sensor. SPIE Proc 5586:136–143
35. Campbell DP, Hartman NF, Walsh JL, Akki U (1994) Integrated optic gaseous NH_3 sensor for agricultural applications. Proc SPIE 2345:314–323
36. Campbell DP, Moore JL, Cobb JM, Edwards JG (1998) Integrated optic sensor for pH and ammonia. Proc SPIE 3537B:319–327
37. Giuliani JF, Wohltjen H, Jarvis NL (1983) Reversible optical waveguide sensor for ammonia vapors. Opt Lett 8:54–56
38. Munir IZ, Hu S, Dordick JS (2002) Chemoenzymatic synthesis of trinitrobenzyl halides as an alternative approach to hexanitrostilbene. Adv Synth Catal 344:1097

Broadband Spectroelectrochemical Interrogation of Molecular Thin Films by Single-Mode Electro-Active Integrated Optical Waveguides

Sergio B. Mendes, S. Scott Saavedra, and Neal R. Armstrong

Abstract Electron transfer processes to/from monolayers or submonolayers of surface-confined molecules are at the core of several established or emerging sensor technologies. Spectroelectrochemical techniques to monitor these redox processes combine spectroscopic information with the normally monitored electrochemical parameters, such as changes in current or voltage, and can be much more sensitive to changes in optical properties coupled with electron transfer than electrochemical techniques alone. Spectroelectrochemical techniques based on absorbance measurements typically suffer from low sensitivity owing to the low concentrations of redox active species on the surface, and their low absorptivities. Electro-active, single-mode waveguide technologies, developed over the last decade, have provided more than adequate sensitivity to characterize electron transfer to surface-confined molecules where the coverage can be as low as a few percent of a monolayer. In this chapter, we review the major developments in combining electrochemical analysis with optical platforms that maximize optical sensitivity, through the development of electro-active integrated planar waveguides operating in the single-mode optical regime. We provide here a general overview of the theoretical formalisms associated with light propagation and absorbance measurements in integrated optical waveguides, and their electro-active counterparts. We also describe the major implementations of the technology, including the extension of the single-mode configuration into a broadband spectroscopic tool to facilitate the interrogation of the entire visible wavelength region during the redox event, and review some specific applications of these techniques, which demonstrate its sensitivity and broad utility.

S.B. Mendes (✉)
Department of Physics and Astronomy, University of Louisville, KY, USA
e-mail: sb.mendes@louisville.edu

S.S. Saavedra and N.R. Armstrong
Department of Chemistry, University of Arizona, AZ, USA

M. Zourob and A. Lakhtakia (eds.), *Optical Guided-wave Chemical and Biosensors I*, Springer Series on Chemical Sensors and Biosensors 7,
DOI 10.1007/978-3-540-88242-8_4,

Keywords Spectroelectrochemistry · Integrated optical waveguides · Electrochemistry · Optical absorbance

Contents

Abbreviations

ASE	Amplified spontaneous emission
ATR	Attenuated total reflectance
CCD	Charge coupled device
CV	Cyclic voltamogramm
Cyt c	Cytochrome c
EA	Electro-active
IOW	Integrated optical waveguide
ITO	Indium tin oxide
MB	Methylene blue
NA	Numerical aperture
S	Sensitivity
TE	Transverse electric
TM	Transverse magnetic

1 Introduction

The need to combine conventional electrochemical measurements of electron transfer (ET) with spectroscopic probes was recognized over 40 years ago [1, 2]. It was quickly realized that optical spectroscopies, first applied to the visible wavelength region, could provide much needed information about the extent and rates of ET, without the complications which accompany the measurement of current or potential versus time, arising from the charging and discharging of the electrical solution double layer adjacent to an electrode surface. Kuwana and coworkers [1, 2] are widely given credit for the development of the first effective spectroelectrochemical methods of electroanalysis, first using transmission through a semi-transparent electrode (the first electrodes were typically glass-coated with 100–200-nm-thick, antimony-doped tin oxide films), and subsequently attenuated total reflectance (ATR) using the same electrodes, in a simple total internal reflection geometry that achieved about 5 to 10 reflections over a total length of several centimeters.

For all such spectroelectrochemical experiments in the visible regime of the electromagnetic spectrum, if at least one redox state is strongly colored, spectroelectrochemical methods can provide a unique probe of the progress of the ET reaction, and characterization of the chemical reactions accompanying the primary ET event. One can reconstruct the Faradaic electrochemical current/voltage response from the absorbance changes accompanying ET, without the current or voltage response due to charging/discharging of the electrochemical double layer (ion motion in the solution region adjacent to the electrode), which accompany all ET reactions at solid electrodes. For low concentrations of the molecule in question, this proves to be a real asset, since the Faradaic current may be nearly impossible to resolve above the charging/discharging background current, and the optical response is completely decoupled from the charging/discharging event. Furthermore, only redox processes that produce an optical change are detected; thus, oxidation/reduction of a chromophore can be selectively detected in the presence of other redox-active but transparent species. An additional feature of spectroelectrochemical interrogation is that it can be performed over a broad spectral bandwidth, which allows for structural characterization of surface-confined chromophores (e.g., dimerization of phthalocyanines is accompanied by a shift in their visible absorbance spectra).

Early on, it was well recognized that spectroelectrochemical spectroscopies could potentially be used to study ET processes of many different types of surface-confined redox active molecules, such as organic dyes and heme proteins. However, many of these molecular systems do not possess a suitable combination of high surface coverage and molar absorptivity, necessitating enhancements in the sensitivity of the spectroelectrochemical platform. The obvious first choice is to decrease the thickness of the ATR element; for example, increasing the number of internal reflections from 1–2 per cm to 10–100 per cm can be achieved by using a 150-μm thick glass coverslip in place of a several millimeter thick ATR element. However, it was eventually realized that moving to single-mode waveguide platforms, where the effective number of reflections could be increased to well over 1,000 per cm, would push the sensitivity of spectroelectrochemical techniques in the visible wavelength region to the point where even submonolayer surface coverages of redox active molecules, with quite low absorptivities, could be characterized.

As a result of the work summarized in this chapter, unprecedented spectroelectrochemical sensitivity is now available in the visible wavelength region using broadband, single-mode waveguide platforms overcoated with semi-transparent conductive oxide electrode layers [3–5]. For certain molecular systems, it is now possible to monitor ET events for surface coverages of just a few percent of a monolayer (~10^{-12} moles/cm^2). This work has spawned the development of broadband spectroelectrochemical sensor platforms, which can combine both electrochemical signals (changes in potential or current versus time) and optical signals, to provide high sensitivities, and in some cases, unprecedented selectivity for multiple analytes whose redox activities are accompanied by spectral changes in distinct wavelength regions.

2 Optical Absorbance in Integrated Optical Waveguides

2.1 Single-Layer Waveguides

A basic waveguide structure, which is sketched in Fig. 1, is composed of a guiding layer surrounded by two semi-infinite media of lower refractive indices. The optical properties of the structure are described by the waveguiding layer refractive index n_w and thickness t, and by the refractive indices of the two surrounding semi-infinite media, here called n_c (for cover) and n_s (for substrate). Application of Maxwell's equations and boundary conditions leads to the well-known waveguide dispersion equation [6]:

$$\frac{2\pi t}{\lambda}\sqrt{n_w{}^2 - N_{eff}{}^2} = m\pi + \tan^{-1}\left[\left(\frac{n_w}{n_c}\right)^{2\rho}\sqrt{\frac{N_{eff}{}^2 - n_c{}^2}{n_w{}^2 - N_{eff}{}^2}}\right] + \tan^{-1}\left[\left(\frac{n_w}{n_s}\right)^{2\rho}\sqrt{\frac{N_{eff}{}^2 - n_s{}^2}{n_w{}^2 - N_{eff}{}^2}}\right], \quad (1)$$

where λ is the light wavelength in vacuum; N_{eff} is the effective refractive index of the waveguide; m is the order of the waveguide mode; $\rho = 0$ applies to transverse electric (TE) modes and $\rho = 1$ applies to transverse magnetic (TM) modes.

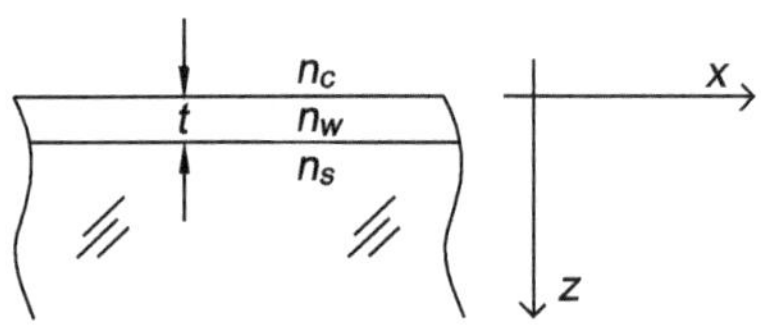

Fig. 1 A basic waveguide structure with a light-confining film of refractive index n_w and thickness t surrounded by two semi-infinite media of lower refractive indices: n_c and n_s

The waveguide effective refractive index, N_{eff}, of each guided mode, at a particular order and polarization, determines the phase velocity (c/N_{eff}) that a particular waveguide mode propagates along the waveguide surface. Its numerical value is typically calculated by solving the transcendental equation (1) using an iterative technique. Once the value of N_{eff} has been determined, then the field profile of that particular mode can be calculated. For instance, for the TE polarization, the electric field is described by [7]:

$$E_y(z) = E_c \exp\left[\frac{2\pi z}{\lambda}\sqrt{N_{eff}^2 - n_c^2}\right], \quad z \leq 0 \tag{2}$$

$$E_y(z) = E_w \cos\left[\frac{2\pi z}{\lambda}\sqrt{n_w^2 - N_{eff}^2} - tan^{-1}\left(\frac{\sqrt{N_{eff}^2 - n_c^2}}{\sqrt{n_w^2 - N_{eff}^2}}\right)\right], \quad 0 \leq z \leq t, \tag{3}$$

$$E_y(z) = E_s \exp\left[-\frac{2\pi(z-t)}{\lambda}\sqrt{N_{eff}^2 - n_s^2}\right], \quad z \geq t, \tag{4}$$

with the following relation between the maximum field amplitudes in each region:

$$E_w\sqrt{n_w^2 - N_{eff}^2} = E_s\sqrt{n_w^2 - n_s^2} = E_c\sqrt{n_w^2 - n_c^2}. \tag{5}$$

As expected for a guided mode, the electric field in the substrate and cover regions decays exponentially, and it shows an oscillatory behavior inside the waveguide film. For the TM polarization, one obtains similar, although not identical, relations for the magnetic field H_y, which then allows the calculation of the Cartesian components E_x and E_z of the electric field using the generalized Ampere's law in the Maxwell's equations [7].

Absorbance calculations in the above configuration are implemented by introducing the absorbing species as a small perturbation to the initial configuration (see Fig. 2). Either a small extinction coefficient is introduced to the optical properties of the cover material (as shown in Fig. 3) or a thin absorbing layer is introduced between the waveguide and the cover (Fig. 4). A critical assumption for the

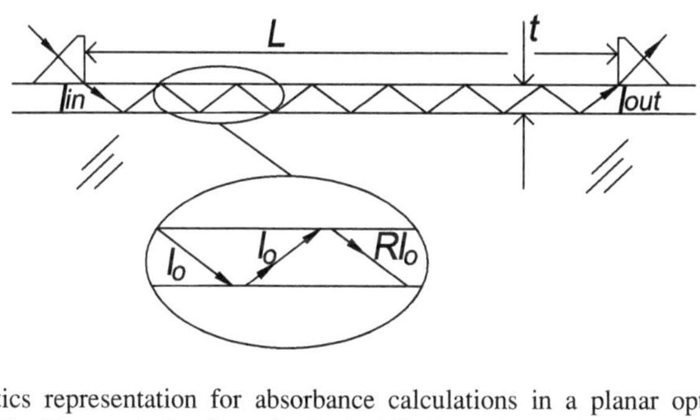

Fig. 2 Ray optics representation for absorbance calculations in a planar optical waveguide. Adapted from [8] with permission from the Optical Society of America, copyright 2000

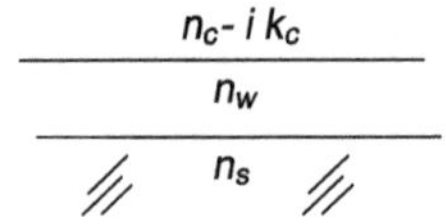

Fig. 3 Bulk solution absorption. Adapted from [8] with permission from the Optical Society of America, copyright 2000

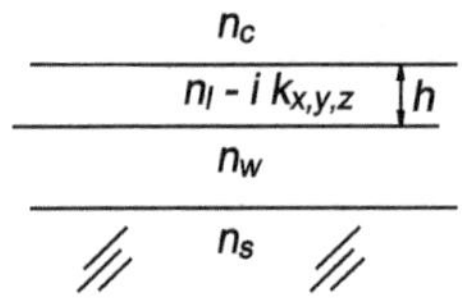

Fig. 4 Absorbance by a surface-adsorbed dichroic layer. Adapted from [8] with permission from the Optical Society of America, copyright 2000

absorbance calculations here is that the perturbation is small enough [8] so that $N_{\rm eff}$ of each mode can be used in the equations to follow. This assumption is valid to most relevant problems to be studied through waveguide modes and does not represent a serious limitation [8]. Under this assumption, the calculation of the attenuation coefficient, or absorbance, is straightforward. The introduction of an absorbing medium in proximity to the waveguide mode frustrates the total internal reflection at the waveguide-cover interface, as indicated at the upper interface in Fig. 2. The attenuation of the propagating guided mode can then be described by the reflectance, R, at the waveguide-cover interface. If we define η as the total number of reflections at the waveguide-cover interface, then the absorbance can be written as [8]:

$$A = -\log_{10}(R^{\eta}) \;\cong\; \frac{\eta\,(1-R)}{\ln(10)}, \tag{6}$$

where we have used the weak perturbation assumption ($R \cong 1$). The total number of reflections along the distance L between the input and output couplers is calculated from the ray optics model and given by:

$$\eta = \frac{\sqrt{n_{\rm w}^2 - N_{\rm eff}^2}}{2N_{\rm eff}}\,\frac{L}{t_{\rm eff}}. \tag{7}$$

The effective thickness, $t_{\rm eff}$, takes into account the Goos-Hänchen shift effects and is given by:

$$t_{\rm eff,\,TE} = t + \frac{\lambda/2\pi}{\sqrt{N_{\rm eff,TE}^2 - n_{\rm c}^2}} + \frac{\lambda/2\pi}{\sqrt{N_{\rm eff,TE}^2 - n_{\rm s}^2}} \tag{8}$$

for the TE polarization, and

$$t_{\rm eff,\,TM} = t + \frac{\lambda/2\pi}{q_{\rm c}\sqrt{N_{\rm eff,TM}^2 - n_{\rm c}^2}} + \frac{\lambda/2\pi}{q_{\rm s}\sqrt{N_{\rm eff,TM}^2 - n_{\rm s}^2}} \tag{9}$$

for the TM polarization, with

$$q_{\mathrm{c,s}} \equiv \left(\frac{N_{\mathrm{eff,TM}}}{n_{\mathrm{w}}}\right)^2 + \left(\frac{N_{\mathrm{eff,TM}}}{n_{\mathrm{c,s}}}\right)^2 - 1. \quad (10)$$

At this point, the reflectance, R, is the sole variable that remains to be determined for the absorbance calculation in (6). The specific calculations of R, and thus A, for the case of bulk solution absorption or surface-immobilized chromophores are described below.

2.1.1 Absorbance from Bulk Solution Species

Figure 3 illustrates a planar waveguide with an absorbing cover medium; light absorption in the cover medium is described by introducing an imaginary term k_c (known as extinction coefficient) to the refractive index. By using the Fresnel reflection coefficient at the waveguide-cover interface and considering the previous assumption of a weakly absorbing medium, $(k_c/n_c) \ll 1$, the reflectance, R, for each polarization is calculated. After those results are inserted into (6), the expression for the absorbance, A_b, as measured through a guided mode probing an absorbing bulk solution, is given by [8]

$$A_{\mathrm{b}} = \varepsilon_{\mathrm{b}}\, c_{\mathrm{b}} \frac{\lambda}{4\pi \left(N_{\mathrm{eff,TE}}^2 - n_{\mathrm{c}}^2\right)^{1/2}} \left\{ \frac{2\, n_{\mathrm{c}} \left(n_{\mathrm{w}}^2 - N_{\mathrm{eff,TE}}^2\right)}{t_{\mathrm{eff,TE}}\, N_{\mathrm{eff,TE}} \left(n_{\mathrm{w}}^2 - n_{\mathrm{c}}^2\right)} L \right\} \quad (11)$$

for TE modes and

$$A_{\mathrm{b}} = \varepsilon_{\mathrm{b}}\, c_{\mathrm{b}} \frac{\lambda}{4\,\pi \sqrt{N_{\mathrm{eff,TM}}^2 - n_{\mathrm{c}}^2}} \left\{ \frac{2\, n_{\mathrm{c}}\, n_{\mathrm{w}}^2 \left(n_{\mathrm{w}}^2 - N_{\mathrm{eff,TM}}^2\right)\left(2\, N_{\mathrm{eff,TM}}^2 - n_{\mathrm{c}}^2\right)}{t_{\mathrm{eff,TM}}\, N_{\mathrm{eff,TM}} \left[n_{\mathrm{w}}^4 \left(N_{\mathrm{eff,TM}}^2 - n_{\mathrm{c}}^2\right) + n_{\mathrm{c}}^4 \left(n_{\mathrm{w}}^2 - N_{\mathrm{eff,TM}}^2\right)\right]} L \right\} \quad (12)$$

for TM modes. In (11) and (12), we have made use of the following relation:

$$\varepsilon_{\mathrm{b}}\, c_{\mathrm{b}} = \frac{4\,\pi\, k_{\mathrm{c}}}{\lambda\, \ln(10)} \quad (13)$$

where ε_{b} is the molar absorptivity of the chromophore and c_{b} is its solution concentration. Equation (13) relates the notations typically used in chemistry and in optics.

2.1.2 Absorbance from Surface-Immobilized Anisotropic and Isotropic Chromophores

Next, we consider the absorbance due to a dichroic adlayer adsorbed onto the waveguide surface with the optical constants as indicated in Fig. 4. The optical properties of the dichroic layer are described by the different extinction coefficients k_x, k_y, and k_z in each Cartesian direction. The reflectance of the waveguide-adlayer-cover system follows the analysis found in Macleod [9] with the anisotropic coefficients taken from Horowitz and Mendes [10]. By assuming a thin and weakly absorbing adlayer, the following expressions are obtained for the absorbance as measured through a guided mode at each polarization [8]:

$$A_l = \varepsilon_l \, \Gamma_l \left\{ \frac{2\, n_l f_y \left(n_w^2 - N_{\mathrm{eff,TE}}^2\right)}{t_{\mathrm{eff,TE}} N_{\mathrm{eff,TE}} \left(n_w^2 - n_c^2\right)} \, L \right\} \tag{14}$$

for TE modes, and

$$A_l = \varepsilon_l \, \Gamma_l \left\{ \frac{2\, n_l\, n_w^2 \left(n_w^2 - N_{\mathrm{eff,TM}}^2\right) \left[\left(N_{\mathrm{eff,TM}}^2 - n_c^2\right) f_x + (n_c/n_l)^4 N_{\mathrm{eff,TM}}^2 f_z \right]}{t_{\mathrm{eff,TM}} N_{\mathrm{eff,TM}} \left[n_w^4 \left(N_{\mathrm{eff,TM}}^2 - n_c^2\right) + n_c^4 \left(n_w^2 - N_{\mathrm{eff,TM}}^2\right) \right]} \, L \right\} \tag{15}$$

for TM modes. In (14) and (15), the adsorbed layer is characterized by the surface coverage Γ_l of the chromophores and the molar absorptivity ε_l of a randomly oriented ensemble of chromophores. We also have used the auxiliary variables f_α defined as $f_\alpha \equiv \frac{k_\alpha}{k_l}$ for $\alpha = $ x,y,z, which is the ratio of the extinction coefficient along a coordinate axis, k_α, and the extinction coefficient of a randomly oriented ensemble of chromophores, k_l. This definition leads to the following relation:

$$\varepsilon_l \, \Gamma_l f_\alpha = \frac{4\pi h}{\lambda \ln(10)} \, k_\alpha. \tag{16}$$

In the isotropic limit, we have $k_\alpha = k_l$ and $f_\alpha = 1$ for $\alpha = x, y, z$. By inserting those values into (14) and (15), we get the expressions for absorbance for the particular case of an isotropic absorbing film.

It is worth noting that, at least in the isotropic limit, the term $\varepsilon_l \, \Gamma_l$ corresponds to the absorbance one would measure by a direct transmission experiment. Therefore, the term inside the bracket in (14) (for TE modes) and (15) (for TM modes) corresponds to the enhancement factor of a long interaction length as provided by the waveguide; we called the term inside the bracket a sensitivity factor (S), which is dependent among other variables on the wavelength, waveguide mode, polarization, and on the distance between the input and output couplers, L. Later in this chapter, we show results of the sensitivity factor per unit length, S/L, for a typical waveguide configuration.

It is also important to consider when both bulk solution species and adsorbed species are simultaneously present in a waveguide measurement. In this case, the total measured absorbance will have a contribution from each term:

$$A = A_{\mathrm{b}} + A_{\mathrm{l}}. \tag{17}$$

The foregoing equations, in the isotropic limit, allow us to write the ratio of the absorbance values for each contribution as:

$$\frac{A_{\mathrm{l}}}{A_{\mathrm{b}}} = \frac{\varepsilon_{\mathrm{l}}\,\Gamma_{\mathrm{l}}}{\varepsilon_{\mathrm{b}}\,c_{\mathrm{b}}\,\frac{\lambda}{4\pi\sqrt{N_{\mathrm{eff}}^2 - n_{\mathrm{c}}^2}}}. \tag{18}$$

When the adsorption process does not significantly change the molecular molar absorptivity (in other words, $\varepsilon_{\mathrm{l}} \cong \varepsilon_{\mathrm{b}}$), then:

$$\frac{A_{\mathrm{l}}}{A_{\mathrm{b}}} \cong \frac{\Gamma_{\mathrm{l}}}{c_{\mathrm{b}}}\,\frac{4\pi\sqrt{N_{\mathrm{eff}}^2 - n_{\mathrm{c}}^2}}{\lambda} = K_{\mathrm{ads}}\,\frac{4\pi\sqrt{N_{\mathrm{eff}}^2 - n_{\mathrm{c}}^2}}{\lambda}, \tag{19}$$

where K_{ads} is the equilibrium constant between the surface-adsorbed and the bulk-dissolved species. By inserting the values of K_{ads} and λ, one can determine if the contribution of bulk species to the overall absorbance measurement is negligible or not.

2.2 *Generalization to Multilayer Integrated Optical Waveguides*

An electro-active planar optical waveguide (even in the single-mode regime) is typically a multilayer structure composed of a layer whose primary function is to confine the light beam and an additional conductive layer to provide an electrochemically active surface. Therefore, the usual three-medium configuration (one layer surrounded by two semi-infinite media) is insufficient for the analysis of electro-active waveguides, so we describe next a methodology more suitable to the general scenario. The approach taken here follows the work of Offersgaard [11]. For completeness and aiming to simplify his formalism for the application in hand, the following equations describe the major steps of his technique. For detailed information, the reader should refer to the original article. In general terms, the goal here is to implement a rigorous formalism where all the layers involved in the waveguide structure are included from the beginning and to solve for the bound modes in an arbitrary multilayer, dichroic, and birefrigent waveguide structure. Each layer is described by the complex refractive index n_{x}, n_{y}, and n_{z} along each Cartesian component, and the layer thickness $d^j \equiv z^{(j)} - z^{(j-1)}$. For each allowed mode and polarization, the effective refractive index of the overall structure and the associated field profile across the structure are the key pieces of information to be

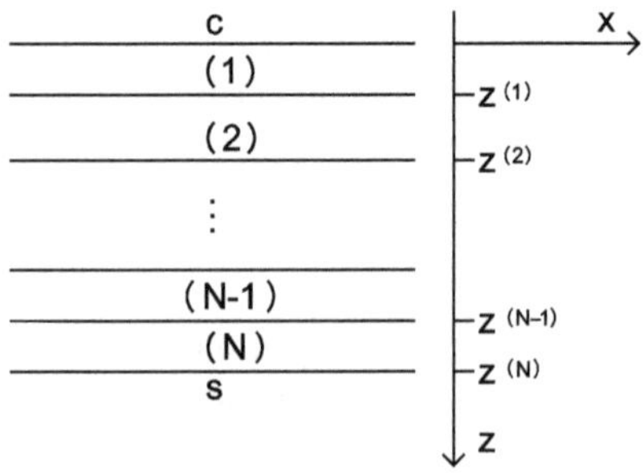

Fig. 5 Coordinate system and multilayer representation

obtained. Once in hand, the imaginary part of the effective refractive index will then determine the attenuation (absorbance) of the propagating beam of the particular waveguide mode.

In Fig. 5, we indicate the coordinate system and the notation used to model the waveguide configuration. The electric and magnetic fields are decomposed into two traveling waves: a positive ($\mathbf{E}_1$, $\mathbf{H}_1$) and a negative ($\mathbf{E}_2$, $\mathbf{H}_2$) propagating wave along the z-direction. Thus, for each medium j we write:

$$\mathbf{E}^{(j)}(\mathbf{r},t) = \mathbf{E}_1^{(j)}(\mathbf{r},t) + \mathbf{E}_2^{(j)}(\mathbf{r},t), \tag{20}$$

$$\mathbf{H}^{(j)}(\mathbf{r},t) = \mathbf{H}_1^{(j)}(\mathbf{r},t) + \mathbf{H}_2^{(j)}(\mathbf{r},t). \tag{21}$$

Each traveling component is described by:

$$\mathbf{E}_{1,2}^{(j)}(\mathbf{r},t) = \frac{1}{2}\,\psi_{1,2}^{(j)}\,\mathbf{p}_{1,2}^{(j)}\,\exp\left\{ i\left[\mathbf{k}_{1,2}^{(j)}.(\mathbf{r} - z^{(j)}\mathbf{e}_z) - \omega\, t\right]\right\} + c.c., \tag{22}$$

$$\mathbf{H}_{1,2}^{(j)}(\mathbf{r},t) = \frac{1}{2}\,\psi_{1,2}^{(j)}\,\mathbf{q}_{1,2}^{(j)}\,\exp\left\{ i\left[\mathbf{k}_{1,2}^{(j)}.(\mathbf{r} - z^{(j)}\mathbf{e}_z) - \omega\, t\right]\right\} + c.c., \tag{23}$$

where $c.c.$ represents the complex conjugate, ψ represents the wave amplitude, $\mathbf{p}$ and $\mathbf{q}$ are the polarization vectors, $\mathbf{k}$ is the wave vector, ω is the optical angular frequency, and $\mathbf{e}_z$ is a unit vector along the z-axis. The Faraday equation $\nabla \times \mathbf{E} = -\mu_0 \frac{\partial \mathbf{H}}{\partial t}$ relates the polarization vectors $\mathbf{p}$ and $\mathbf{q}$ through the wave vector $\mathbf{k}$:

$$\mathbf{q}_{1,2}^{(j)} = \frac{\mathbf{k}_{1,2}^{(j)} \times \mathbf{p}_{1,2}^{(j)}}{\omega\,\mu_0}, \tag{24}$$

where μ_0 is the permeability of free space. For light propagation in the x–z plane, the wave vectors may be written as:

$$\mathbf{k}_1^{(j)} \equiv \frac{2\pi}{\lambda}\begin{pmatrix}\beta\\ 0\\ \alpha^{(j)}\end{pmatrix}, \tag{25}$$

$$\mathbf{k}_2^{(j)} \equiv \frac{2\pi}{\lambda} \begin{pmatrix} \beta \\ 0 \\ -\alpha^{(j)} \end{pmatrix}, \tag{26}$$

where λ is the light wavelength in vacuum. From the boundary conditions for the tangential field components, we get:

$$\begin{pmatrix} \psi_1^{(c)} \\ \psi_2^{(c)} \end{pmatrix} = \mathbf{M} \cdot \begin{pmatrix} \psi_1^{(s)} \\ \psi_2^{(s)} \end{pmatrix}, \tag{27}$$

with:

$$\mathbf{M} \equiv (\mathbf{D}^{(c)})^{-1} \left(\prod_{j=1}^{N} \mathbf{G}^{(j)} \right) (\mathbf{D}^{(s)}), \tag{28}$$

$$\mathbf{G}^{(j)} \equiv \begin{pmatrix} \cos\left(\frac{2\pi}{\lambda}\alpha^{(j)} d^{(j)}\right) & -i\frac{p_{t1}^{(j)}}{q_{t1}^{(j)}} \sin\left(\frac{2\pi}{\lambda}\alpha^{(j)} d^{(j)}\right) \\ -i\frac{q_{t1}^{(j)}}{p_{t1}^{(j)}} \sin\left(\frac{2\pi}{\lambda}\alpha^{(j)} d^{(j)}\right) & \cos\left(\frac{2\pi}{\lambda}\alpha^{(j)} d^{(j)}\right) \end{pmatrix}, \tag{29}$$

$$\mathbf{D}^{(j)} \equiv \begin{pmatrix} p_{t1}^{(j)} & p_{t2}^{(j)} \\ q_{t1}^{(j)} & q_{t2}^{(j)} \end{pmatrix}. \tag{30}$$

In (29) and (30), the subscript t refers to the tangential components of the corresponding vector.

For the TE polarization, the tangential component of the polarization vectors is given by:

$$p_{y1} = p_{y2} = 1, \tag{31}$$

and application of (24) gives us:

$$q_{x1} = -q_{x2} = -\sqrt{\frac{\varepsilon_0}{\mu_0}}\,\alpha, \tag{32}$$

where ε_0 is the permittivity of free space.

As previously described, we assume a diagonal tensor for the permittivity

$$\frac{\varepsilon}{\varepsilon_o} = \begin{pmatrix} n_x^2 & 0 & 0 \\ 0 & n_y^2 & 0 \\ 0 & 0 & n_z^2 \end{pmatrix}, \tag{33}$$

which can describe most of the experimental configurations of interest. By inserting (33) into the wave equation, we get

$$\left[\frac{\varepsilon}{\varepsilon_0} - \left(\frac{c}{\omega}\right)^2 |\mathbf{k}|^2\right]\mathbf{E} + \left(\frac{c}{\omega}\right)^2 (\mathbf{k} \cdot \mathbf{E})\,\mathbf{k} = 0, \tag{34}$$

so we can relate both components of the wave vector by the dispersion equation:

$$\alpha = \sqrt{n_y^2 - \beta^2}, \tag{35}$$

where the root, to provide decaying fields far from the guide, should be chosen according to:

$$\mathrm{Im}\,(\alpha) > 0. \tag{36}$$

For the TM polarization, we find the tangential components of the polarization vector are given by:

$$p_{x1} = -p_{x2} = \frac{\alpha}{\sqrt{\left|\beta\,(n_x/n_z)^2\right|^2 + |\alpha|^2}}, \tag{37}$$

and

$$q_{y1} = q_{y2} = -\sqrt{\frac{\varepsilon_0}{\mu_0}}\frac{(n_x)^2}{\sqrt{\left|\beta\,(n_x/n_z)^2\right|^2 + |\alpha|^2}}. \tag{38}$$

Again, the wave equation relates the components of the wave vector by:

$$\alpha = \frac{n_x}{n_z}\sqrt{n_z^2 - \beta^2}, \tag{39}$$

where the root should also be chosen according to (36).

The determination of β, the only remaining unknown, is provided by solving

$$m_{11}(\omega, \beta) = 0 \tag{40}$$

to produce a bound mode in the waveguide region; m_{11} is an element of the matrix defined in (28). After denoting the solution to (40) as

$$\beta_0 = N_{\mathrm{Re}} - i\,N_{\mathrm{Im}}, \tag{41}$$

the absorbance is calculated as:

$$A = \frac{4\pi L}{\lambda \ln 10} N_{\mathrm{Im}}. \tag{42}$$

2.3 Sensitivity of Single- and Multimode Waveguide Structures

Operation of planar waveguides in the single-mode regime offers several advantages when compared with multimode configurations in applications of absorbance spectroscopy. Most importantly, the sensitivity for probing chromophores in proximity of the waveguide surface is far superior in the single-mode structure compared with the multimode structure. As a metric to quantify this point, Fig. 6 shows the sensitivity factor per unit length, S/L, defined as the absorbance per unit length, A_l/L, measured through a guided mode propagating for a distance L inside the waveguide divided by the absorbance measurement in direct transmission configuration, $\varepsilon_l \Gamma_l$, for probing a surface-adsorbed thin film. For convenience, the data are plotted with the V-number as the x-axis, which is related to the waveguide thickness by: $V = \frac{2\pi t\sqrt{n_w^2 - n_s^2}}{\lambda}$. For small thicknesses (small V-number), the waveguide operates in the single-mode regime and the sensitivity reaches a peak at approximately $V = 1.53$ for TE and $V = 1.73$ for TM. Below the peak, as V-number decreases and N_{eff} approaches n_s, the Goos-Hanchen shift effect dominates the effective thickness and makes the guided modes excessively large, decreasing the sensitivity. As the thickness increases beyond the ideal V-number, the sensitivity of the lowest-order mode decreases rapidly. At the same time, new modes are allowed to propagate in

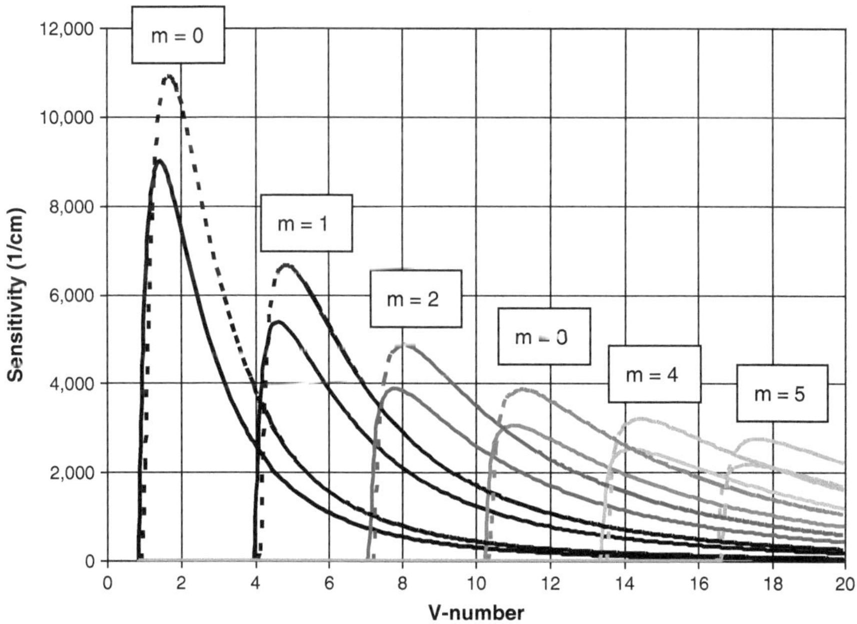

Fig. 6 Sensitivity factor per unit length for absorbance measurement through propagating guided modes. Refractive index profile is given by 1.33 (water cover), 1.56 (waveguide film; e.g. Corning glass 7059), and 1.46 (fused silica substrate). *Solid lines*: TE modes; *dashed lines*: TM modes. Wavelength for calculations: $\lambda = 550$ nm

the guiding structure; however, their sensitivity is much less than that in the single-mode regime.

Another complicating factor in multimode waveguides is that the sensitivity needs to be evaluated by the weighted average of all the modes that propagate in the structure; the weighting factor measures the fraction of optical power coupled into each mode. Therefore, in order to retrieve any information from measurements with a multimode waveguide, a precise knowledge of the power distribution at the input coupling port is required. An additional complication for multimode waveguides is that power hopping from mode to mode is always present to some degree because of scattering mechanisms along the propagation length of the guide, and this can also disturb the retrieved information.

2.4 Polarized Measurements and Molecular Orientation

Electromagnetic radiation of any frequency is described by the interplay of oscillating electric and magnetic fields, which are vectors and therefore require a direction in space for a full description. The optical absorption of an electromagnetic wave by matter is determined by specifying the electric field vector associated with the electromagnetic radiation and the vector transition dipole for the molecule under investigation. As expressed in the Fermi's Golden rule,

$$A = \left\langle \left|\vec{\mu}.\vec{E}\right|^2 \right\rangle. \tag{43}$$

By controlling the orientation in space of the probing electric field, one can gain information on the orientation of the transition dipoles being probed by the light beam. Usually, that information is described as an order parameter, which is intrinsically an ensemble average over the population of molecules interacting with the light beam. If one then knows the relationship between the transition dipole with respect to the molecular structure (e.g., the transition dipole angle with respect to the molecular plane of a polyaromatic hydrocarbon) it is possible to infer the molecular orientation in space.

Optical modes of different polarizations (TE or TM) in planar waveguides interact differently with chromophores located in the proximity to the waveguide surface. Typically, the electric fields associated with TE and TM modes have different strengths at the interface and different profiles across the waveguide structure, and those features lead to different optical interaction with the absorbing species even when those species are isotropically (or randomly) distributed in space. Equations (14) and (15) provide a quantitative description of those effects and it is common to say that the pathlength for TE and TM are different. For a simple and well-defined waveguide structure (e.g., the step-index single-layer waveguide), it is possible to derive analytical expressions as described in Sect. 2.2. In the case more of complex structures (e.g., multilayer and/or gradient-index), a complete and accurate description of the waveguide for predicting the behavior of TE and TM can be difficult. A simplifying approach that is particularly useful in these difficult cases, and is also applicable in general, is to initially calibrate the planar waveguide with an isotropic probe, which can be either an absorbing species dissolved in the bulk phase or a thin-film that is known to have random dipole orientation when immobilized on the waveguide. By measuring the absorbance at both polarizations for such an isotropic probe, we can factor out the difference in pathlength that is due to solely to the difference in the electric fields of the TE and TM modes. With such calibration in hand, one can then apply the calibrated waveguide to samples of interest. As previously described [12], the normalized dichroic ratio defined by

$$\rho_{\text{norm}} \equiv \frac{\rho_{\text{sample}}}{\rho_{\text{iso}}} = \frac{\left(A_{\text{TE}}/A_{\text{TM}}\right)_{\text{sample}}}{\left(A_{\text{TE}}/A_{\text{TM}}\right)_{\text{iso}}} \tag{44}$$

can be related to the dipole components along the Cartesian axes using

$$\rho_{\text{norm}} = \frac{\left\langle \mu_y^2 \right\rangle \left(2N_{\text{eff,TM}}^2 - n_c^2\right)}{\left\langle \mu_x^2 \right\rangle \left(N_{\text{eff,TM}}^2 - n_c^2\right) + \left\langle \mu_z^2 \right\rangle N_{\text{eff,TM}}^2}, \tag{45}$$

with a numerical value bound as follows:

$$0 \leq \rho_{\text{norm}} \leq \frac{2N_{\text{eff,TM}}^2 - n_c^2}{N_{\text{eff,TM}}^2 - n_c^2}. \tag{46}$$

From the normalized dichroic ratio, one can calculate the dipole component along each Cartesian axis (in = x, y; out = z) as:

$$\frac{\left\langle \mu_{\text{in}}^2 \right\rangle}{\mu^2} = \frac{1}{2} - \frac{\left\langle \mu_{\text{out}}^2 \right\rangle}{2\,\mu^2} = \frac{\rho_{\text{norm}}\, N_{\text{eff,TM}}^2}{\left(2N_{\text{eff,TM}}^2 - n_c^2\right) + \rho_{\text{norm}}\left(N_{\text{eff,TM}}^2 + n_c^2\right)}, \tag{47}$$

from which the order parameter associated with the one photon process (absorption) can be calculated by [13]:

$$\langle P_2(\theta) \rangle = \frac{3\left\langle \mu_{\text{out}}^2 \right\rangle}{2\,\mu^2} - \frac{1}{2} \tag{48}$$

For a more comprehensive description of the molecular orientation, higher-order parameters are typically needed to add independent information for the reconstruction of a probability distribution function [14]. The ability to detect molecular orientation changes during ET events of surface-confined molecules is an extremely powerful capability afforded by the use of electro-active waveguides. It has often been suspected that addition or subtraction of an electron from an isolated molecule on a surface would be accompanied by, and controlled by, an orientation change of that molecule.

3 Instrumental Setup

Optical spectroscopy with single-mode planar waveguide requires bright light sources (high power per mode), sensitive detectors, and efficient waveguide couplers. Furthermore, these optical components need to perform well over a broad spectral band. Current detector technologies, either single-channel detectors (e.g., avalanche photodiode, photomultiplier) or array detectors (CCD) are generally suitable. Input and output optical waveguide couplers capable of handling broadband light sources that are spatially incoherent (composed of a large number of spatial modes) with limited brightness are certainly a major challenge for the implementation of broadband spectroscopy with single-mode planar optical waveguides for interrogation of molecular films.

3.1 Waveguide Couplers

Input and output couplers for single-mode planar optical waveguides can be broadly classified as end-facet couplers or transverse couplers. End-facet couplers work quite well in particular for channel waveguides with cross-sectional dimensions of a few microns in height and width [15]. However, for waveguides with submicron dimensions in the transverse direction, the end-facet approach is problematic as it requires the formation of a smooth and optically flat facet right at the very end of the waveguide device. In those cases, transverse couplers such as prism and grating couplers are usually the preferred choice.

3.1.1 Prism Coupler

A prism in close proximity to a waveguide film can be used to excite waveguide modes [16] as long as the prism's refractive index is higher than the effective index of the particular guided mode. As shown in Fig. 7, the effective refractive index of the coupler (also know as Snell invariant, $N_{\mathrm{p}} = n_{\mathrm{p}} \sin\theta_{\mathrm{p}}$, and defined as the projection of the k-vector of the incoming beam onto the waveguide surface divided by $2\,\pi/\lambda$) is given by:

$$N_{\mathrm{p}} = n_{\mathrm{p}} \sin\theta_{\mathrm{p}} = n_{\mathrm{i}} \sin\theta_{\mathrm{i}} \cos\varphi + \sqrt{n_{\mathrm{p}}{}^2 - n_{\mathrm{i}}{}^2 \sin^2\theta_{\mathrm{i}}}\, \sin\varphi, \tag{49}$$

Fig. 7 Input prism coupler

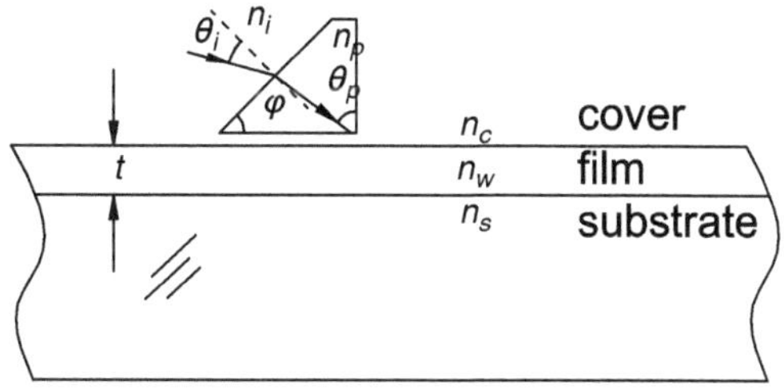

where φ is the prism base angle. By tuning the angle θ_i of the incident beam, one can adjust the coupler effective index N_p to match the effective refractive index of the waveguide, i.e.,

$$N_p = N_{eff}. \tag{50}$$

As previously discussed, N_{eff} is found for each mode and polarization by solving either (1) in the case of a single-layer waveguide or (40) for the more complex configurations.

The effective refractive indices of the waveguide $N_{eff}(\lambda)$ and the prism coupler $N_p(\lambda)$ depend on the wavelength. In the case of the prism coupler, the dispersion of $N_p(\lambda)$ comes from the material dispersion as expressed in (49); in the case of the waveguide effective refractive index, the material dispersion is also important but an additional major contribution to the overall dispersion comes from the mode confinement (known as modal dispersion) which can be seen by the explicit wavelength dependence of (1). Typically, for a particular angle of incidence of the incoming broadband beam, only a center wavelength and a small band around it will give an efficient match between the coupler and the waveguide effective refractive index. In general, either the angle needs to be tuned to couple different wavelengths or an angular width (also known as numerical aperture, NA) in the optical beam needs to be provided to achieve broadband operation. An incident beam with an appropriate angular width was the approach taken by Kato et al. [17] and Bradshaw et al. [3, 18] to achieve broadband coupling in single-mode planar optical waveguides. Prisms with a proper dispersive power can also be designed to match the dispersion behavior between coupler and waveguide, and therefore significantly reduce the angular width needed to couple a broader spectral range [19].

3.1.2 Grating Coupler

The effective refractive index, N_g, of a diffraction grating coupler, as illustrated in Fig. 8, is described by [20–22]:

$$N_g = n_i \sin\theta_i + m\,\frac{\lambda}{\Lambda}, \tag{51}$$

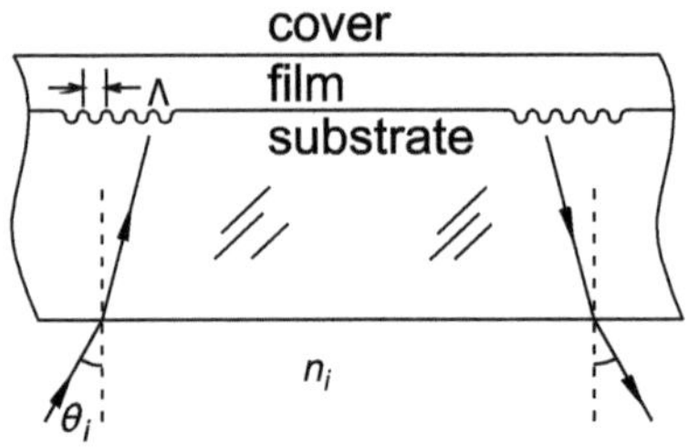

Fig. 8 Input and output grating couplers

where the first term in the right-hand side accounts for the Snell invariant, Λ is the grating period, and m is the period of the grating. As before, the effective refractive index of the coupler needs to match that of the waveguide mode, i.e.,

$$N_g = N_{\text{eff}}, \tag{52}$$

which is done by tuning the angle θ_i for each wavelength. Unlike the prism coupler, the grating coupler can be implemented with the beam illumination either from the top side or from the bottom side of the device.

Grating couplers are typically integrated into the waveguide device and buried underneath the waveguide layer by first creating a periodic corrugation on the surface of the substrate slide and then depositing the waveguide layer. This process of creating a surface corrugation in the substrate usually involves the fabrication of a holographic pattern, done in a Lloyd's mirror configuration, to imprint an interference pattern in a photoresist film deposited on the substrate. The photoresist film is developed to create a photopattern with a periodic modulation. Dry etching of the photopattern is then employed to transfer the modulation to the substrate slide. After removal of any remaining photoresist, the samples with the periodic corrugation are then overcoated with the waveguide structure of choice [23].

Unlike the prism coupler which requires precise alignment of an extra external optical element to excite a guided mode, grating couplers are fully embedded in the optical device and simplify tremendously its incorporation into a spectroscopic instrument. As mentioned above, both grating and prism couplers require the adjustment of the incoming optical beam to the proper angle to match the effective index of the coupler and the waveguide; in addition, both couplers require the lateral position of the optical beam to be located close to the edge of the coupler to achieve strong coupling efficiencies [24].

Comparing the prism coupler to the grating coupler, the former has the advantage of low dispersion and therefore requires a smaller NA for achieving broadband coupling. An inconvenience, however, is the requirement of mounting the prism in close proximity to the waveguiding layer, separated by a very small (in the wavelength range) and precisely fixed gap. Changes in the gap affect the coupling efficiency substantially and perturb the measurements. Although grating couplers are fully integrated optical components that provide strong robustness to the coupling process, they require either a scanning angle capability or a very high NA to achieve broadband operation.

3.1.3 Achromatic Coupler

An approach that aims to overcome the difficulties above involves the combination of a grating coupler with two pre-dispersive components: an additional grating and a prism to minimize the overall mismatch between the coupler and waveguide effective indices [25]. The additional grating is designed to approximately cancel the dispersion of the grating coupler; the prism is added solely to improve the cancellation of the effective index mismatch and is optically attached to the back side of the device without requiring stringent control of the gap. In this case, broadband operation with single-mode waveguides was achieved and demonstrated for submonolayer protein films [26].

3.2 Broadband Light Sources and Lasers

Let us now consider spectroscopic measurements over a broad spectral band, such as that emitted by a blackbody source. A typical tungsten–halogen lamp operated at a temperature of 3,200 K in the tungsten filament (with an emissivity of 0.33) has an emission spectrum with a power density of about −75 dBm/nm per mode (or 33 pW/nm per mode) around the center of the visible spectrum (550 nm). Such low brightness, expressed either by the optical power/(unit wavelength × mode) or by optical power/(unit wavelength × emitting area × solid angle), identifies a major challenge for using typical broadband light sources with planar waveguides operating in the single-mode regime at the transverse direction of the guiding structure.

The maximum power that can be coupled into each mode of a waveguide structure is given by the power per mode emitted by the light source being deployed. Passive optical components (without a gain medium as in lasers or optical amplifiers) cannot increase the brightness of an optical beam; the best they can do is sustain the power per mode of an incoming beam. In channel waveguides and optical fibers, with both transverse and lateral confinements, one can calculate the total number of modes for the particular structure and estimate the maximum power that can be coupled into the device. In slab planar waveguides, with confinement only in the transverse direction, one can increase the device throughput by launching several modes in the lateral dimension. Although in this case, the sensitivity for probing surface events is approximately the same for all the lateral modes, an enhanced throughput can be quite helpful in increasing the signal-to-noise ratio of the spectroscopic measurement. In other words, a single-mode slab optical waveguide corresponds to several hundreds (or even thousands) of channel waveguides, all of them probing simultaneously surface-adsorbed species with equal sensitivity.

Regarding the source brightness, tunable lasers are certainly an alternative for overcoming the low brightness of incoherent sources; however, their higher cost can be a limiting factor for several applications. The arrival of GaN LED technologies at shorter wavelengths in the visible spectrum and into the near UV may become a useful alternative for the applications discussed here, as they may provide higher brightness than broadband filament or arc lamp sources. Although LED brightness is lower than that of lasers, they offer broader spectral emission which partially circumvents the need for tunability in laser sources. We also note the emergence of new technologies such as supercontinuum generation and amplified spontaneous emission (ASE) that can potentially offer new alternatives for higher brightness broadband sources to be incorporated into waveguide spectroscopic instrumentation.

When photometric detection with a single wavelength or narrow band is sufficient, spatially coherent (spatially single-mode) laser sources are the preferred choice as they offer high brightness and are easily coupled into single-mode optical waveguides. Because of the high brightness of readily available laser beams and the simplicity in setting up a waveguide coupler for a single wavelength, the majority of applications with single-mode integrated optical waveguides have been limited to single wavelength measurements. However, in many of those cases, acquisition of broadband spectroscopic data would be much preferred to enable molecular structure to be characterized and overlapping spectral signals of multiple chromophores to be resolved.

3.3 Electro-Active Optical Waveguides: Materials and Fabrication

Itoh and Fujishima were among the first to perform both photometric and electrochemical interrogations on an electro-active planar optical-waveguide platform [27, 28]. In their 1988 report, they described a channel gradient-refractive-index glass waveguide fabricated by the ion exchange process and overcoated by spray pyrolysis with an electro-active, antimony-doped, tin oxide layer to provide for both electrochemical and optical detection of surface-adsorbed species. Simultaneous acquisition of the cyclic voltammogram and the corresponding intensity of an outcoupled beam from a propagating guided wave excited with a 633 nm He-Ne laser were obtained for an adsorbed layer of methylene blue (MB). MB is a convenient redox couple with which the sensitivities of thin waveguide platforms with the original ATR-based spectroelectrochemical platforms can be compared. MB adsorbs to oxide surfaces and has a well-known, reversible two-electron reduction at −0.275 V and a high-molar absorptivity for the oxidized form at 633 nm, so that probing its activity with conventional He–Ne lasers is straightforward.

Itoh and Fujishima reported a sensitivity factor for the optical signal of 20–40 for a multimode structure and approximately 150 for a single-mode waveguide [27, 28]. These sensitivity factors are consistent with the weak confinement provided by a gradient-refractive index waveguide. Another relevant work was reported by Piraud et al. on the development of a chemical sensor based on the opto-electrochemical response of an electro-active ion-exchange channel waveguide overcoated with a redox active film [29, 30].

As described above, ion-exchange waveguides certainly provide enhanced sensitivity when compared with optical interrogation in direct transmission geometry. However, their signal enhancement is characteristically less than what can be achieved by a step-refractive index profile. As shown in Fig. 6, a typical step-refractive index single-mode waveguide with submicron optical confinement provides several orders of magnitude (about 9,000/cm for TE and 11,000/cm for TM in the example of Fig. 6) in sensitivity enhancement per centimeter of beam propagation along the device.

To achieve higher sensitivity, Dunphy et al. [4, 5] developed a planar single-mode electro-active waveguide. Although formed by a stack of three layers (each layer with a constant refractive index), the optical device was designed and fabricated to operate in the single-mode regime with tight optical confinement, and thus high sensitivity. As schematically shown in Fig. 9, the multilayer electro-active waveguide structure consisted of Corning 7059 glass, silicon dioxide, and indium tin oxide (ITO) layers that were deposited on a soda lime glass substrate (75mm × 25mm × 1 mm) using a RF sputtering technique. The primary function of the Corning glass layer (refractive index $n = 1.56$, thickness $t = 400$ nm) is to provide for most of the optical confinement. The next layer, SiO_2, with $n = 1.46$ and $t = 200$ nm, functions as a buffer to minimize possible ion migration between the Corning glass layer and the overlying ITO layer; such migration could result in greater optical propagation losses and reduced electrical conductivity in the ITO layer. The ITO layer, with $n = 2.0$ and $t = 50$ nm, was deposited under optimized partial pressure of O_2 to achieve an appropriate balance of optical transparency and electrical conductivity. After RF sputtering deposition, an annealing process was performed at 225°C to improve the ITO conductivity by reducing grain boundary defects. For a 50-nm-thick ITO layer, Dunphy et al. [4, 5] reported resistance values of 700–800 Ω/□.

As illustrated in Fig. 9 and described in Sect. 3.1.2, prior to sputtering deposition of the layers, two surface relief gratings (period of $\Lambda = 400$ nm) were fabricated on the glass slide for input and output coupling of the optical beam. A calculation based on the theory developed in Sect. 2.2 gave an electric field profile for the TE mode as shown in Fig. 10.

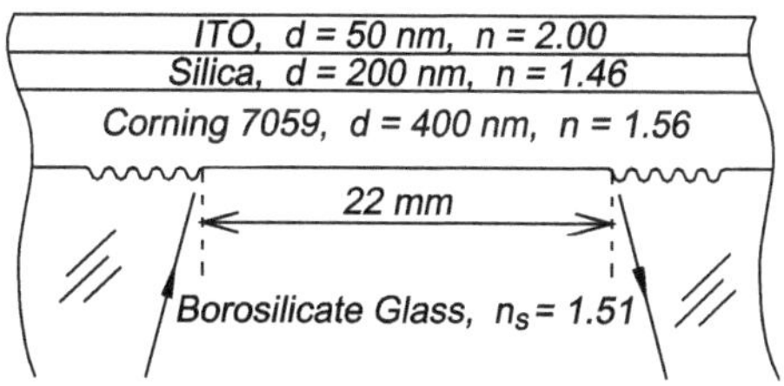

Fig. 9 Electro-active planar single-mode optical waveguide. Adapted from [4] with permission from the America Chemical Society, copyright 1997

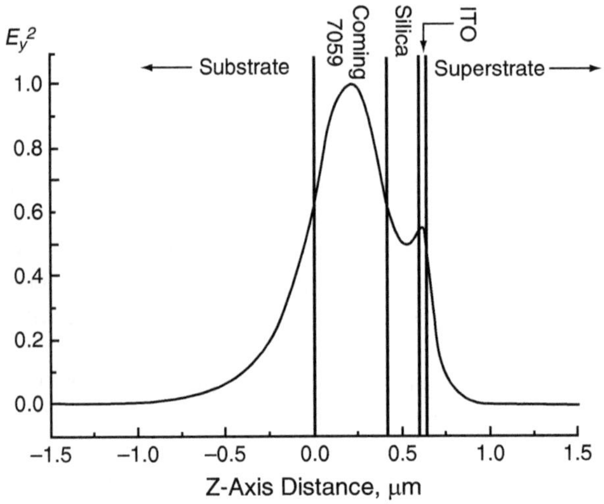

Fig. 10 Electric field profile in TE polarization across the multilayer electro-active waveguide structure. Adapted from [4] with permission from the America Chemical Society, copyright 1997

4 Characterization and Applications

4.1 Adsorbed Dyes

Dunphy et al. [4, 5] and Mendes et al.[4] first applied the electro-active single-mode waveguide (Figs. 9 and 10) for electrochemical characterization of adsorbed MB since exploration of this redox couple provided a direct comparison to the sensitivity of previously developed spectroelectrochemical platforms. Figure 11 summarizes the results. The reduction of MB to its leuco (transparent at 633 nm) form results in transmittance increase in the waveguide; however, this behavior is superimposed on a loss of transmittance arising from background changes in the optical properties of the ITO thin film with increasing negative applied potentials as shown in Fig. 11a. Hansen et al. [1] were the first to describe this phenomenon for spectroelectrochemical platforms, showing it to arise from the changes in electron density of states in the near surface region of the oxide, causing an increased light attenuation in the background signal that is almost linear with applied potential. Fortunately, this background attenuation can be easily subtracted, leaving the absorbance change versus potential shown in Fig. 11b.

The derivative of the absorbance with respect to the applied potential, $\mathrm{d}A/\mathrm{d}E$, allows one to create an absorpto-voltammogram, shown in Fig. 12 (a), where the data for $\mathrm{d}A/\mathrm{d}E$ are plotted versus the electric potential for both the forward (reduction of MB to its leuco form) and reverse (reoxidation of leuco MB to MB) sweeps. These experimental results can be quite useful when we consider that double layer charging usually does not produce major changes in the spectroscopic

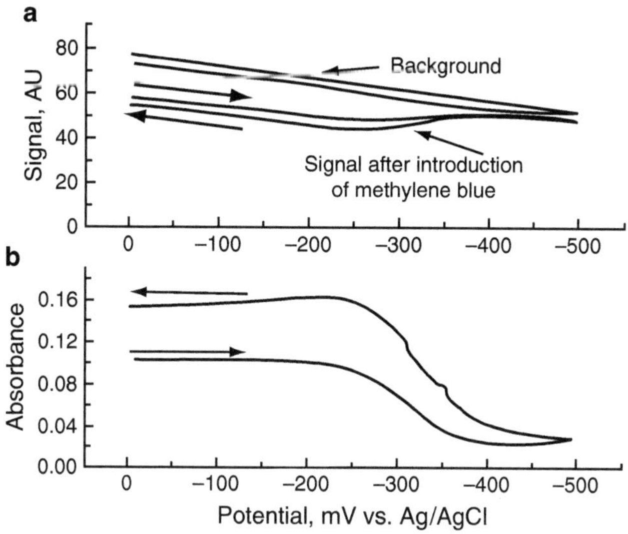

Fig. 11 (**a**) Change in the optical signal as a function of applied potential during reduction (negative going sweep) and oxidation for both solution background and a surface-confined MB submonolayer. There is a significant change in background transmittance of the ITO/waveguide platform as a function of applied potential arising from the change in free electron population in the near surface region as the electrode potential move towards negative potentials; however, these optical changes are easily removed to provide the plot shown in (**b**) for the net absorbance due solely to the redox couple. The two-electron, one-proton reduction process converts the blue form of MB to its leuco form, which is fairly transparent at the working wavelength of 633 nm. Reprinted from [5] with permission from Marcel Dekker, copyright 1999

data (or produces changes that can be easily removed), and under those conditions, the Faradaic current can be related to the optical changes in the adsorbed redox species by [31]:

$$\frac{\mathrm{d}A}{\mathrm{d}E} = \frac{S\left(\varepsilon_{\mathrm{O}} - \varepsilon_{\mathrm{R}}\right)}{nFBv}\, i_{\mathrm{F}}, \tag{53}$$

where n is the electric charge per molecule, F is the Faraday constant, B is the electrode area, v is the scan rate, E is the applied potential, S is the waveguide sensitivity factor, and ε_{O} $(\varepsilon_{\mathrm{R}})$ is the molar absorptivity of the oxidized (reduced) species. As described in (53), the derivative of the optical absorbance with respect to the applied potential is directly proportional to the Faradaic current, with no contribution from the non-Faradaic component. A key advantage of the optical interrogation, when compared to the conventional cyclic voltammetry, is the removal of the non-Faradaic components in the optical signal.

An electro-active integrated optical waveguide (EA-IOW) with a separation of 8 mm between the input and output grating couplers was used to experimentally assess the sensitivity of EA-IOW-based measurements on surface-confined films. Measurements performed on adsorbed MB ($\varepsilon = 7{,}800\ \mathrm{M}^{-1}\ \mathrm{cm}^{-1}$ at 633 nm) at a

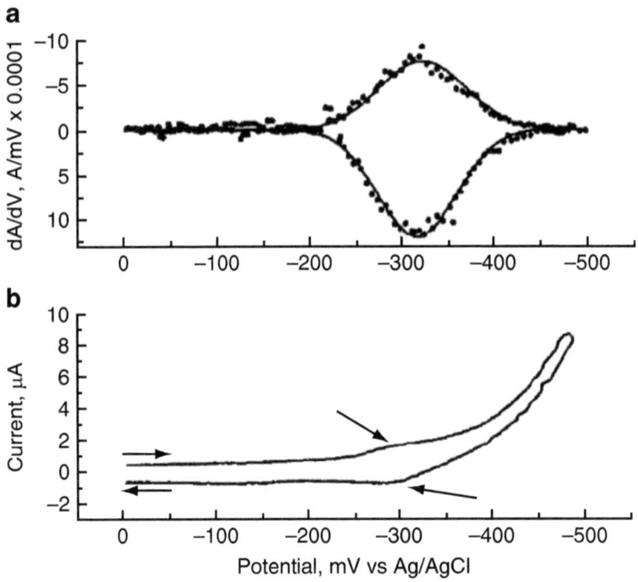

Fig. 12 (**a**) The reconstructed absorpto-voltammogram and (**b**) the original cyclic voltammogram (current versus applied voltage, as voltage is swept continuously to about −0.5 volts and back to 0.0 volts) for surface-confined MB on the ITO-coated single-mode waveguide. The cyclic voltammogram shows a nearly indistinguishable peak on the forward and reverse sweeps corresponding to the reduction and oxidation of surface-confined MB, masked by the background current associated with charging of the electrical double layer at the electrode/solution interface. The absorptovoltammogram is obtained by taking the first derivative of the background-corrected absorbance (Fig. 11) versus applied potential, and shows high contrast during the reduction of MB to the leuco (transparent) form of the dye. There is a larger optical change on the reverse sweep (oxidation of leuco form back to MB) than on the forward sweep (reduction of MB to the leuco form), which we have hypothesized arises from changes in packing density of the leuco form of the dye, once formed, leading to higher overall surface coverages. The data shown were obtained at slow sweep rates; at higher sweep rates, which did not provide time for this rearrangement, the asymmetry in the peak shapes was removed. Reprinted from [5] with permission from Marcel Dekker, copyright 1999

surface coverage of approximately 4% of a close-packed monolayer showed that the EA-IOW is about 4,000 times more sensitive than a single-pass transmission geometry, which represented a significant improvement over earlier electro-active waveguide platforms.

4.2 *Spectroelectrochemistry of Cytochrome c* Films

Due to potential applications in biosensing and to develop a better understanding of heterogeneous biological electron transfer reactions, direct electrochemistry between adsorbed proteins and solid electrodes has been studied extensively [32,

33]. Much of the research in this field has dealt with horse cytochrome *c* (cyt *c*), a small electron transfer protein that contains a redox-active, iron porphyrin prosthetic group [34]. It is widely hypothesized that in order for facile electron transfer to occur between surface-adsorbed cyt *c* and a solid electrode, the protein molecules must be oriented with the heme pocket facing the electrode surface [34, 35]. This favorable structure is thought to result from electrostatic interactions between the asymmetric distribution of positive charges on cyt *c* and the negatively charged electrode surface. Conventional electrochemical techniques are useful for characterizing reaction rates in these systems but cannot provide a rigorous assessment of the oriented adsorption hypothesis because they do not provide structural information.

A comparison of conventional cyclic voltammetry (CV) and EA-IOW-based spectroelectrochemistry of cyt *c* adsorbed to a waveguide surface is shown below. The CV data plotted in Fig. 13 were recorded at several scan rates [36]; even at the highest scan rate, non-Faradaic background overwhelms the Faradaic portion of the signal. An example of an absorpto-voltammogram measured on an equivalent cyt *c* film in TM polarization at 514.5 nm is shown in Fig. 14a. The corresponding reconstructed voltammogram, shown in Fig. 14b (*solid line*), demonstrates the power of EA-IOW-based spectroelectrochemistry to completely eliminate the non-Faradaic background. The optical data can also be measured using TE-polarized light, and the respective reconstructed voltammogram is plotted in Fig. 14b as well (*dotted line*). As discussed above, the ratio of the absorbances in TE and TM provides information about the orientation of the chromophores in the film (in this case, the iron porphyrin in cyt *c*). Thus, the data in Figs. 13 and 14 demonstrate that EA-IOW-based spectroelectrochemistry provides simultaneous information about both electron transfer rates and molecular orientation for ultra-thin adsorbed protein films.

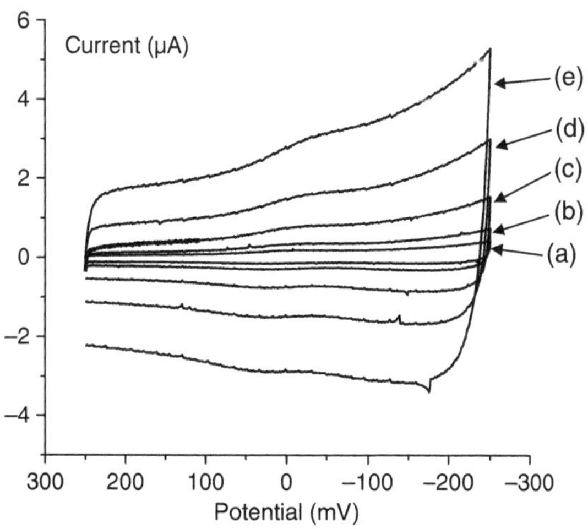

Fig. 13 Cyclic voltammograms of cyt *c* adsorbed to ITO in 5 mM, pH 7 phosphate buffer, at varying scan rates: (**a**) 10, (**b**) 20, (**c**) 50, (**d**) 100, and (**e**) 200 mV/s. The potential is reported versus an Ag/AgCl reference electrode. The protein surface coverage is about 8 pmol/cm^2; [36]

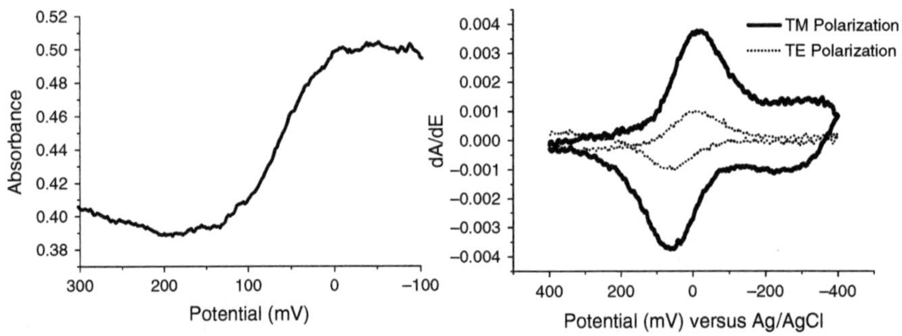

Fig. 14 (**a**) The absorbance response, measured at 514.5 nm in TM polarization, of a film of cyt *c* (about 8 pmol/cm^2) adsorbed to an EA-IOW measured during a potential scan from +400 to −400 mV. (**b**) An optically detected cyclic voltammogram (*solid line*) reconstructed from the data in (**a**). The TE-polarized, optically detected cyclic voltammogram is also shown (*dotted line*) [36]

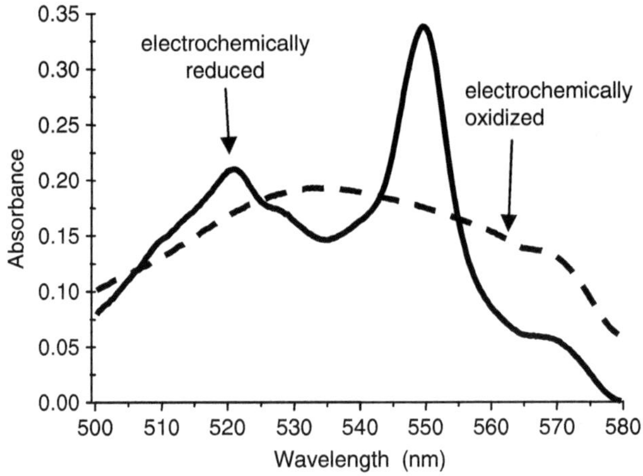

Fig. 15 TM-polarized EA-IOW spectra of a submonolayer of cyt *c* adsorbed to the surface of the EA-IOW: (1) electrochemically reduced ferrocyt *c* (*solid line*), collected at a potential of −400 mV vs. Ag/AgCl; (2) electrochemically oxidized ferricyt *c* (*dashed line*), collected after scanning the potential to +400 mV. Reprinted from [3] with permission from the America Chemical Society, copyright 2003

4.3 Broadband Spectroscopy on an EA-IOW

The studies described in Sects. 4.1 and 4.2 were performed with monochromatic laser sources. To provide the information content of broadband spectroscopy, Bradshaw et al. [3] extended the prism coupling approach of broad angular width described above to create a broadband-coupled EA-IOW. The result is a spectrometer that combines the extreme sensitivity of the single-mode EA-IOW with a multichannel spectroscopic capability over a 200 nm spectral bandwidth.

In Fig. 15, EA-IOW spectra of an adsorbed cyt *c* layer at potentials necessary to maintain the protein in its fully reduced state or fully oxidized state are shown. Scanning to and maintaining the potential at −400 mV vs. Ag/AgCl reduces the cyt *c* layer, resulting in the appearance of the sharp band near 550 nm, which is characteristic of ferrocyt *c*. Scanning to and maintaining the potential at +400 mV oxidizes the protein, producing a spectrum with a single, broad band in the 500–600 nm range, which is characteristic of ferricyt *c*. These spectra were acquired in the TM_0 mode and the entire spectral range from ~500 to ~700 nm was probed simultaneously. Spectra can also be acquired in the TE_0 mode, which allows molecular order parameters to be measured under potential control over a broad spectral bandwidth. This combination of features is unprecedented.

5 Concluding Remarks

The sections above illustrate the combination of sensitivity and high information content that can be provided by the EA-IOW platforms. Future applications will likely include characterization of relationships between molecular structure and charge transport in ultra-thin films. The ability to study submonolayer films has the potential to yield significant insight into current problems in electrochemistry such as hypothesized relationships between molecular packing, orientation, and electron transfer kinetics. Another important application is chemical sensing based on spectroelectrochemical selectivity, e.g., sensors in which absorbance, luminescence, and an electrochemical parameter, such as current or interfacial potential, can be monitored simultaneously to produce greater orthogonality of data in response space relative to standard electrochemically based sensors (e.g., ion-selective electrodes) [37].

Current technical difficulties mostly center on fabrication challenges appropriate to combine high transparency to achieve low propagation loss and high conductivity on the surface chemistry of ITO, which is rough, chemically heterogeneous, and metastable. Instrumental difficulties in light coupling, especially for broadband interrogation with low brightness sources, would benefit from a more user-friendly apparatus to facilitate a wider range of applications.

References

1. Hansen WN, Kuwana T et al (1966) Observation of electrode-solution interface by means of internal reflection spectrometry. Anal Chem 38(13):1810–1821
2. Kuwana T, Heineman WR (1976) Study of electrogenerated reactants using optically transparent electrodes. Acc Chem Res 9(7):241–248

3. Bradshaw JT, Mendes SB et al (2003) Broadband coupling into a single-mode, electroactive integrated optical waveguide for spectroelectrochemical analysis of surface-confined redox couples. Anal Chem 75(5):1080–1088
4. Dunphy DR, Mendes SB et al (1997) The electroactive integrated optical waveguide: ultrasensitive spectroelectrochemistry of submonolayer adsorbates. Anal Chem 69(15):3086–3094
5. Dunphy DR, Mendes SB, et al (1999) Spectroelectrochemistry of monolayer and submonolayer films using an electroactive integrated optical waveguide, Chapter 29. In: Wieckowski A (Ed) Interfacial electrochemistry. Marcel Dekker, New York
6. Tien PK (1971) Light waves in thin films and integrated optics. Appl Opt 10(11):2395–2413
7. Kogelnik H (1979) Theory of dielectric waveguides. Integrated optics, 2nd edn. New York, Springer, pp 13–81
8. Mendes SB, Saavedra SS (2000) Comparative analysis of absorbance calculations for integrated optical waveguide configurations by use of the ray optics model and the electromagnetic wave theory. Appl Opt 39(4):612–621
9. Macleod HA (1986) Thin-film optical filters. Macmillan, New York
10. Horowitz F, Mendes SB (1994) Envelope and wave-guide methods – a comparative-study of PbF2 and CeO2 birefringent films. Appl Opt 33(13):2659–2663
11. Offersgaard JF (1995) Wave-guides formed by multiple layers of dielectric, semiconductor, or metallic media with optical loss and anisotropy. J Opt Soc Am A 12(10):2122–2128
12. Mendes SB, Bradshaw JT et al (2004) Technique for determining the angular orientation of molecules bound to the surface of an arbitrary planar optical waveguide. Appl Opt 43(1):70–78
13. Zannoni C, Arcioni A et al (1983) Fluorescence depolarization in liquid-crystals and membrane bilayers. Chem Phys Lipids 32(3–4):179–250
14. Runge AF, Saavedra SS et al (2006) Combination of polarized TIRF and ATR spectroscopies for determination of the second and fourth order parameters of molecular orientation in thin films and construction of an orientation distribution based on the maximum entropy method. J Phys Chem B 110(13):6721–6731
15. Kremeskötter J, Wilson R et al (1995) Detection of glucose via electrochemiluminescence in a thin-layer cell with a planar optical waveguide. Meas Sci Technol 6:1325–1328
16. Tien PK, Ulrich R (1970) Theory of prism-film coupler and thin-film light guides. J Opt Soc Am 60(10):1325–1337
17. Kato K, Takatsu A et al (1995) A slab-optical-wave-guide absorption-spectroscopy of Langmuir-Blodgett-films with a white-light excitation source. Chem Lett 6:437–438
18. Bradshaw JT, Mendes SB et al (2002) A simplified broadband coupling approach applied to chemically robust sol-gel, planar integrated optical waveguides. Anal Chem 74(8):1751–1759
19. Mendes SB, Li LF et al (1997) Achromatic prism-coupler for planar waveguide. Opt Commun 136(3–4):320–326
20. Dakss ML, Kuhn L et al (1970) Grating coupler for efficient excitation of optical guided waves in thin films. Appl Phys Lett 16(12):523–525
21. Harris JH, Winn RK et al (1972) Theory and design of periodic couplers. Appl Opt 11 (10):2234–2241
22. Kogelnik H, Sosnowski TP (1970) Holographic thin film couplers. Bell Syst Tech J 49(7):1602–1608
23. Li L, Xu M et al (1987) Fabrication of photoresist masks for submicrometer surface relief gratings. Proc SPIE 835:72–82
24. Tamir T (1982) Integrated optics. Springer, Berlin
25. Mendes SB, Li LF et al (1995) 70-nm-bandwidth achromatic wave-guide coupler. Appl Opt 34(27):6180–6186
26. Mendes SB, Li LF et al (1996) Broad-band attenuated total reflection spectroscopy of a hydrated protein film on a single mode planar waveguide. Langmuir 12(14):3374–3376
27. Itoh K, Fujishima A (1988) An application of optical waveguides to electrochemistry: construction of optical waveguide electrodes. J Phys Chem 92:7043–7045

28. Itoh K, Fujishima A (1992) An application of optical waveguides to electrochemical and photoelectrochemical processes. In: Murphy OJ, Srinivasan S, Conway BE (eds) Electrochemistry in transition. Plenum, New York, pp 219–225
29. Piraud C, Mwarania E et al (1992) An optoelectrochemical thin-film chlorine sensor employing evanescent fields on planar optical waveguides. Anal Chem 64:651–655
30. Piraud C, Mwarania EK et al (1992) Optoelectrochemical transduction on planar optical wave-guides. J Lightwave Technol 10(5):693–699
31. Bard AJ, Faulkner LR (1980) Electrochemical methods. Wiley, New York
32. Bowden EF (1997) Wiring mother nature: interfacial electrochemistry of proteins. Electrochem Soc Interface 6:40–44
33. Gorton L, Lindgren A et al (1999) Direct electron transfer between heme-containing enzymes and electrodes as basis for third generation biosensors. Anal Chim Acta 400:91–108
34. Hawkridge FM, Taniguchi I (1995) The direct electron transfer reactions of cytochrome c at electrode surfaces. Comments Inorg Chem 17:163–187
35. Song S, Clark RA et al (1993) Characterization of cytochrome c/alkanethiolate structures prepared by self-assembly on gold. J Phys Chem 97:6564–6572
36. Robertson RT (2002) The development of electroactive total internal reflection optical devices for the characterization of metalloprotein films. PhD dissertation. Department of Chemistry, Tucson, Arizona
37. Ross SE, Seliskar CJ et al (2000) Spectroelectrochemical sensing based on multimode selectivity simultaneously achievable in a single device. 9. Incorporation of planar waveguide technology. Anal Chem 72:5549–5555

Part II
Plasmonic-Waveguide Sensors

Surface Plasmon Resonance: New Biointerface Designs and High-Throughput Affinity Screening

Matthew J. Linman and Quan Jason Cheng

Abstract Surface plasmon resonance (SPR) is a surface optical technique that measures minute changes in refractive index at a metal-coated surface. It has become increasingly popular in the study of biological and chemical analytes because of its label-free measurement feature. In addition, SPR allows for both quantitative and qualitative assessment of binding interactions in real time, making it ideally suited for probing weak interactions that are often difficult to study with other methods. This chapter presents the biosensor development in the last 3 years or so utilizing SPR as the principal analytical technique, along with a concise background of the technique itself. While SPR has demonstrated many advantages, it is a nonselective method and so, building reproducible and functional interfaces is vital to sensing applications. This chapter, therefore, focuses mainly on unique surface chemistries and assay approaches to examine biological interactions with SPR. In addition, SPR imaging for high-throughput screening based on microarrays and novel hyphenated techniques involving the coupling of SPR to other analytical methods is discussed. The chapter concludes with a commentary on the current state of SPR biosensing technology and the general direction of future biosensor research.

Keywords Surface plasmon resonance · Microarray · SPR imaging · Protein-carbohydrate · Protein-lipid · Lectin

Contents

M.J. Linman and Q.J. Cheng (✉)
Department of Chemistry, University of California, Riverside, CA 92521, USA
e-mail: quan.cheng@ucr.edu

M. Zourob and A. Lakhtakia (eds.), *Optical Guided-wave Chemical and Biosensors I*, Springer Series on Chemical Sensors and Biosensors 7,
DOI 10.1007/978-3-540-88242-8_5, © Springer-Verlag Berlin Heidelberg 2010

Abbreviations

SPR	Surface plasmon resonance
k_{ass}	Association constant
k_{diss}	Dissociation constant
HEG	Hexaethylene glycol spacer
SNA	*Sambucus nigra agglutinin*
HMGA-2	High-mobility-group transcriptional factor
smGFM	Soluble green fluorescent protein
CaM	Calmodulin
K_D	Equilibrium dissociation constant
K_A	Equilibrium association constant
SELEX	Systematic evolution of ligands by exponential enrichment
ELISA	Enzyme-linked immunosorbent assay
IE	Imaging ellipsometry
CBPs	Carbohydrate-binding proteins
tBLM	Tethered bilayer membrane
GM1	Monosialotetrahexosylganglioside
GC	Gas chromatography
VEGF	Vascular endothelial growth factor
SPFS	Surface plasmon fluorescence spectroscopy
PNAs	Peptide nucleic acids
k_{light}	Photon wave vector
RU	Resonance units
SPRi	Surface plasmon resonance imaging
A	Analyte
SPs	Surface plasmons
E	Evanescent field
ERα	Estrogen receptor α
MEL	Mannosylerythritol lipid

RBP4	Retinol binding protein 4
ssDNA	Single-stranded DNA
GNP	Gold nanoparticle
PDMS	Poly(dimethylsiloxane)
LTP	Lipid transfer protein
IgG	Human immunoglobulin G
LC	Liquid chromatography
MS	Mass spectrometry
HRP	Horseradish peroxidase
TOF	Time-of-flight
GAG	Glycosaminoglycan
k_{sp}	Surface plasmon wave vector

1 Introduction to Surface Plasmon Resonance

Surface plasmon resonance or SPR is a surface-sensitive analytical technique that is widely used to monitor both chemical and biological interactions in real time. The real-time aspect of this optical technique gives SPR a distinct advantage over endpoint binding assays, which are limited and have been gradually replaced by SPR since its inception in the 1990s. The phenomenon of surface plasmon resonance occurs at a metal/dielectric interface where one of the biological binding pairs is immobilized on the metal surface while the other binding partner is allowed to flow across the sensing interface. SPR spectroscopy monitors the changes in refractive index occurring at the metal surface upon interaction between the two bio-specific ligands. All analyses require no labels, thus precluding the use of convoluted, and sometimes disruptive, labeling chemistry found in fluorescence methods. Given this perceived analytical advantage, the use of surface plasmon resonance has grown substantially in the recent years. According to a recent survey by SciFinder Scholar, from the year 2007 to the present (May 2008), over 2,000 articles and reviews have been published on surface plasmon resonance analysis, including one comprehensive review published in early 2008 [1].

Despite the obvious advantages of a label-free technique, SPR is a nonselective detection because anything that binds to the surface will change the refractive index. Therefore, reproducible and well-understood surface chemistry to create a selective sensing interface is an important research area for understanding the nature of biological interactions. This chapter covers in detail the principles of SPR from a physical standpoint and then examines unique surface chemistry and assay approaches for various biological and chemical interactions. In addition, SPR imaging for high-throughput screening based on microarrays and emerging hyphenated techniques involving the coupling of SPR to other analytical methods is discussed. Finally, the chapter concludes with a commentary on the current state of SPR sensing technology and the general direction of future biosensor research.

2 SPR: Physical Aspects and Kinetics of Sensing

2.1 Theory

Surface plasmons (SPs) are surface electromagnetic waves that propagate parallel along a metal/dielectric interface. For this phenomenon to occur, the real part of the dielectric constant of the metal must be negative, and its magnitude must be greater than that of the dielectric. Thus, only certain metals such as gold, silver, and aluminum are usually used for SPR measurements. The dispersion relation for surface plasmons on a metal surface is:

$$k_{\mathrm{sp}} = \frac{w}{c}\left(\frac{\varepsilon(\omega)\varepsilon_{\mathrm{a}}}{\varepsilon_{\mathrm{a}} + \varepsilon(\omega)}\right)^{1/2}, \qquad (1)$$

where k_{sp} is the wave vector for surface plasmons, c is the speed of light, $\varepsilon(\omega)$ is the metal's complex dielectric function, and ε_{a} is the dielectric constant of the ambient [2]. Given this dispersion relation, surface plasmons can be directly excited by electrons but not directly by light because SPs have a longer wave vector than light waves of the same energy ($k_{\mathrm{light}} = \omega/c$) [3]. Thus, the wave vector of the photon must be increased in order to convert the energy of the photon into surface plasmons. This can be done by employing a high refractive index prism or grating coupler. There are two major setups to excite surface plasmons: the Otto configuration and the Kretschmann configuration. Since the Otto configuration is rarely employed, we focus our discussion entirely on the Kretschmann configuration. This configuration (shown in Fig. 1) employs p-polarized light that is totally internally reflected at the metal surface. The matching conditions for the wave vectors of the incident light and SP can be achieved by tuning either the incident angle or the

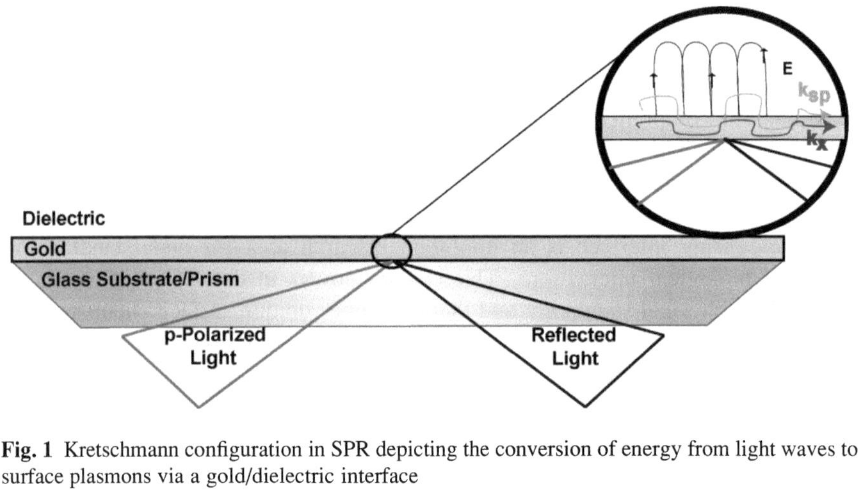

Fig. 1 Kretschmann configuration in SPR depicting the conversion of energy from light waves to surface plasmons via a gold/dielectric interface

wavelength of the incident light. Using a fixed wavelength light source, surface plasmons will be generated in the metal film at angles where the photon's wave vector (k_{light}) equals the surface plasmon's wave vector (k_{sp}). Since this phenomenon is a conversion of light energy to surface plasmons, the excitation of SPs corresponds to an attenuation of the reflected light intensity. The angle where a complete attenuation of the reflected light occurs is known as the surface plasmon resonance angle and is depicted as a symmetric dip in the reflection spectrum.

Along with the creation of SPs, there exists a decaying evanescent wave, depicted as **e** in Fig. 1. The evanescent field (E) associated with the surface plasmon wave has its maximum in the surface and decays exponentially into the space perpendicular to it, extending into the metal and the dielectric [4–6]. The position of the resonance angle is affected by the binding of biomolecules to the metal surface. Consequently, SPR is sensitive to changes in the surface characteristics near the interface at a distance of about 200 nm [7], and thus offers a great potential as a surface analytical technique for label-free and nondestructive study of interfacial properties and processes involving both chemical and biological species.

2.2 *Instrumentation and Modes of Operation*

SPR instrumentation has been commercialized by a number of companies, including Biacore (now GE Healthcare), Sensata Technologies (formerly part of Texas Instruments), Eco Chemie, Biosensing Instrument, and Biosuplar, to name a few. Since the Biacore instrument has proven to be the most popular for SPR chemical and biological sensors [8], a brief discussion is merited. The sensing chip in the Biacore instrument makes contact with a microfluidic system to create several flow cells through which solutions are independently passed under the control of valves. The back side of the chip couples with the optical system, in which light is reflected at a range of angles of incidence from the chip onto a detector that enables the resonance angle to be measured and logged by the control software as a function of time [9]. The angle is conventionally expressed in terms of the arbitrary resonance units, RU, where a change in resonance angle of 0.1° is equal to 1,000 RU [10]. The Biacore instrument is fully automated, highly sensitive, and requires only a minute amount of samples for analysis. However, the cost of the instrument is usually high and the use of prefabricated sensor chips also drives up the running cost. Nevertheless, it has been the top choice for biologists whose goals are to characterize binding affinity rather than method development. With the growing industry of SPR technology, there is a plethora of options and instruments available to suit various research needs.

It should be noted that there exist angular-scanning SPR and wavelength-scanning SPR, though angular-scanning is more widely practiced. In the Kretschmann configuration in the angular scanning mode, there is a light source (e.g., a laser) that is polarized upon hitting a high-refractive index prism under total internal reflection conditions (Fig. 2). The SPR signal is detected by monitoring the reflected light intensity on the opposite side of the prism, which changes upon an analyte binding

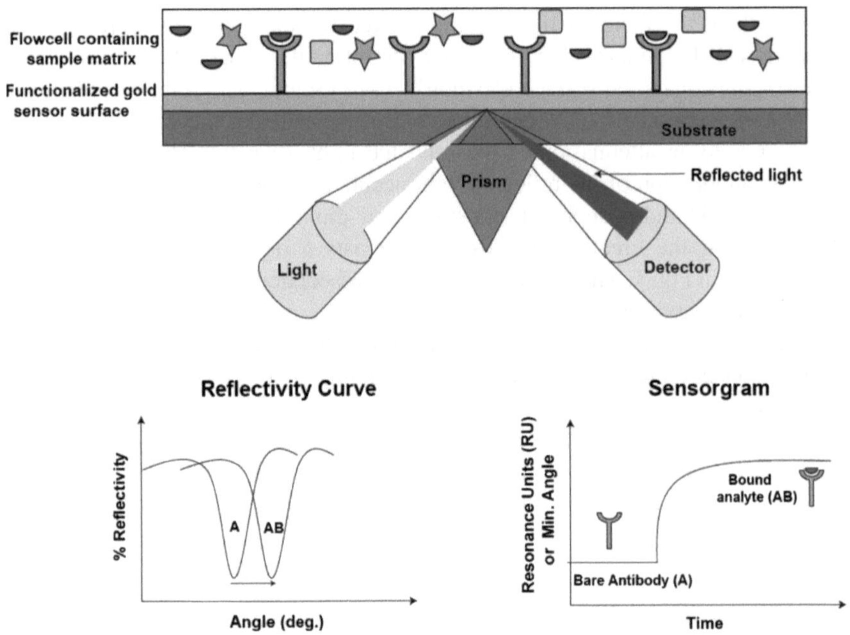

Fig. 2 Schematic of real-time biointerface upon analyte binding to sensing surface. The reflectivity curve (*lower left*) shifts upon analyte binding (displayed as AB) and can be monitored in real-time with a sensorgram (*lower right*)

to the sensing surface. The binding curve obtained by monitoring the minimum angle shift as a function of time is known as a sensorgram, shown in the bottom right hand portion of Fig. 2, which results from a shift in minimum angle in the reflectivity curve shown on the left side of the same figure. The analysis of SPR via sensorgrams is discussed in the following section.

2.3 Sensorgrams and Kinetics

The SPR sensorgram generally contains three phases: the association phase, the dissociation phase, and the regeneration phase, as shown in detail in Fig. 3. The binding kinetics that quantitatively characterizes a bio-molecular interaction by rate constants and equilibrium constants can be determined from the sensorgram.

For an SPR measurement, the reaction rate and equilibrium constants of interactions can be assessed with the reaction:

$$A + B \rightarrow AB, \tag{2}$$

where A is the biological/chemical analyte and B is the ligand immobilized on the gold sensor surface. The association rate is the rate at which complex AB forms.

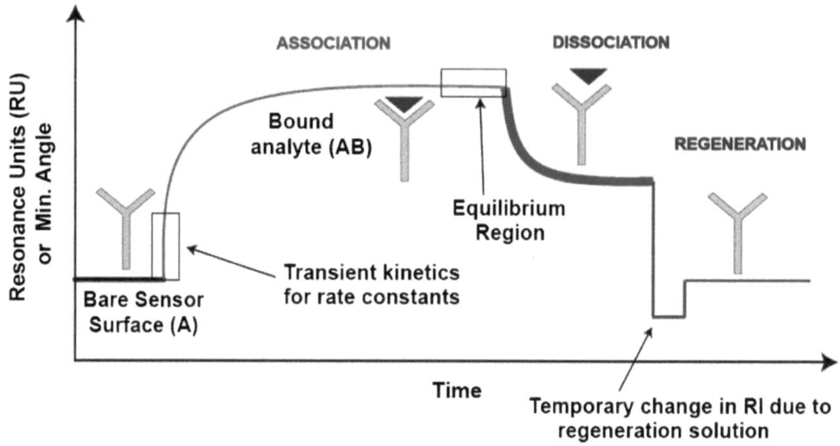

Fig. 3 Sample SPR sensorgram showing all the different kinetics phases along with key functional regions for data analysis

Given that binding can occur very quickly, the transient kinetic period is important for determining the association constant and the methods to determine its value have been demonstrated previously [11]. Generally, in this initial rate kinetic method, at $t = 0$, the equation for initial rate analysis is:

$$\frac{dAB}{dt} = AB_{\max}[A]k_{\text{ass}}, \quad (3)$$

where $AB_{\max}$ is the maximum response that can be obtained for analyte binding on the sensor surface, and k_{ass} is the association constant in units of $\text{mol}^{-1}\text{s}^{-1}$. By plotting the initial rate against analyte concentration A, a straight line is obtained with a slope equal to $AB_{\max} \times k_{\text{ass}}$. After k_{ass} is determined, k_{diss}, the dissociation constant, in units of s^{-1} for an AB-type reaction, can be determined mathematically by (4):

$$AB_t = (AB_0 - AB_\infty)[\exp(-k_{\text{diss}}t)] + AB_\infty, \quad (4)$$

where AB_0 is the initial response (i.e., the beginning of the dissociation curve), and AB_∞ is the final response once completely dissociated.

As the formation of AB complexes reaches equilibrium, the equilibrium association constant and the equilibrium dissociation constant can be determined. These constants represent affinities of interaction rather than kinetic constant values. Briefly, the value for the equilibrium dissociation constant, K_D, can be determined from the rate constants by the equation:

$$K_D = \frac{k_{\text{diss}}}{k_{\text{ass}}}, \quad (5)$$

where the reciprocal of K_D is the equilibrium association constant, K_A. In addition, the equilibrium dissociation constant can be determined from the sensorgram by:

$$AB_{eq} = AB_{max}(1/(1 + K_D/[A])), \tag{6}$$

where AB_{eq} is the average of the response signal at equilibrium in defined intervals for each concentration of analyte $[A]$[12]. It should be noted that the kinetics methods displayed here are the most commonly employed. There are other methods that are constantly being employed for SPR for specific biological interactions [13,14].

2.4 SPR Imaging

SPR imaging (SPRi) is a related technique to SPR spectroscopy but utilizes the fixed angle measurement of changes in reflectivity across the sensing surface rather than the angular shift. The basic components of an SPR imaging system and corresponding images are shown in Fig. 4. Upon analyte binding to array elements, a shift in the SPR minimum angle leads to a change in percent reflectivity. The change can be monitored in a quantitative and visual fashion as difference images (Fig. 4). A major difference between SPR and SPRi in instrumental setup is that SPRi typically uses a CCD camera for image collection and processing. SPRi allows for quantitative determination of analyte presence and is also much more amenable to high-throughput screening. The recent advancement in this area is covered later in the chapter. For more information about SPR imaging, readers are advised to consult some recent reviews on the subject [3, 15].

3 Use of SPR for Bio-interaction Analysis

The principal application of SPR is affinity analysis in biological systems. We focus on bio-interaction analysis with SPR in three general categories: carbohydrate–protein, DNA–protein/protein–protein, and lipid–protein interactions. These studies make up the major areas of SPR-based biological sensors.

3.1 Protein–Carbohydrate Interactions

Protein–carbohydrate interactions are important in cellular signaling and cancer cell metastasis [16, 17]. The ability to monitor the interactions in a quantitative,

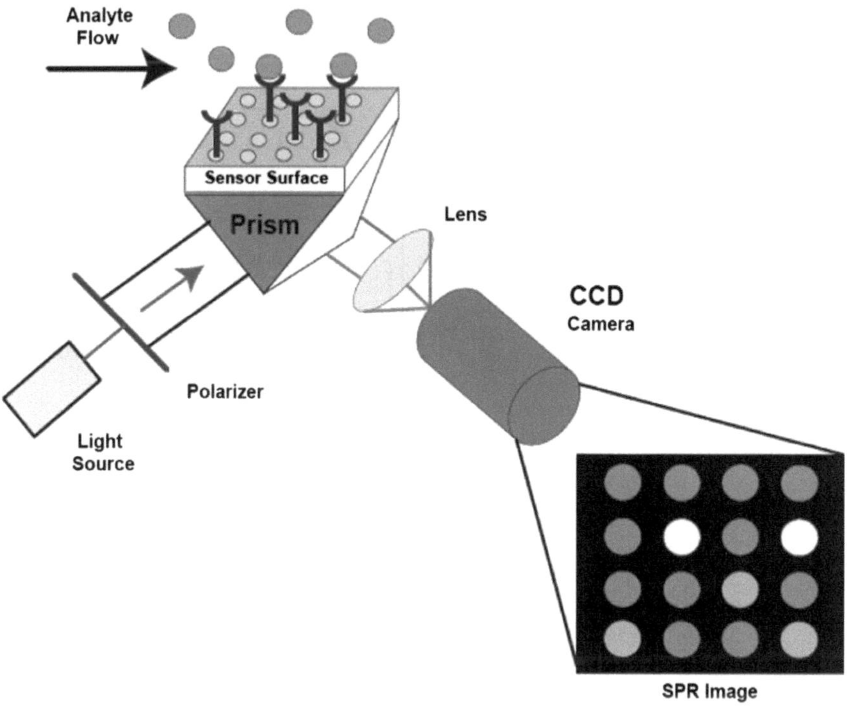

Fig. 4 A schematic SPR imaging setup. The brighter the array element, the more material present at that location

real-time, and label-free manner makes SPR ideally suited for the work, and it has been the focus of many research groups, including ours [18]. We recently reported the fabrication of a novel sensing interface of biotinylated sialosides to probe lectin–carbohydrate interactions using SPR [18]. The tethering of synthesized carbohydrates to a gold surface using biotin–NeutrAvidin interactions and the implementation of an inert hydrophilic hexaethylene glycol spacer (HEG) between the biotin and the carbohydrate resulted in a well-defined interface, enabling desired orientational flexibility and enhanced access of binding partners. The specificity of lectin binding was demonstrated with nanomolar sensitivity. This system could illuminate small differences in carbohydrate structure on the basis of SPR signal. Figure 5 shows an SPR sensorgram for the binding of *Sambucus nigra agglutinin* (SNA) to a carbohydrate-functionalized surface. As evidenced from Fig. 5, this surface design enables multiple experiments to be performed on the same substrate using a glycine stripping buffer, which selectively regenerates the surface without damaging the sensing interface. In addition, we demonstrated a quantitative comparison of binding for different carbohydrates surfaces to saturation concentrations of SNA (Fig. 6). With only small differences in structure (either a different functional group off the C5 carbon in sialic acid or a different sialyl linkage), SPR data unequivocally discriminate the carbohydrates in a real-time

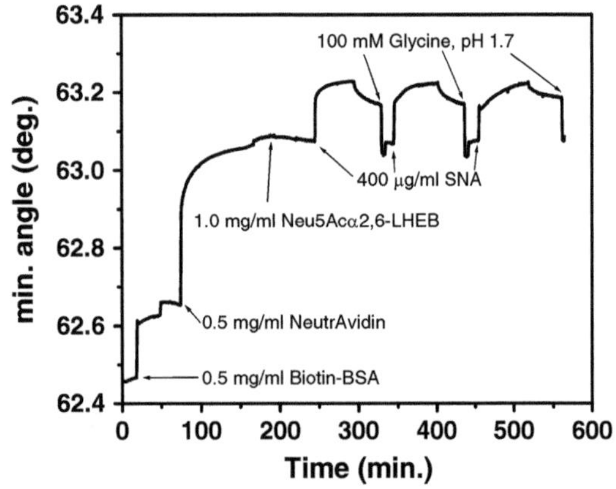

Fig. 5 Characteristic sensorgram for carbohydrate functionalized sensing surface in response to 400 μg/mL SNA

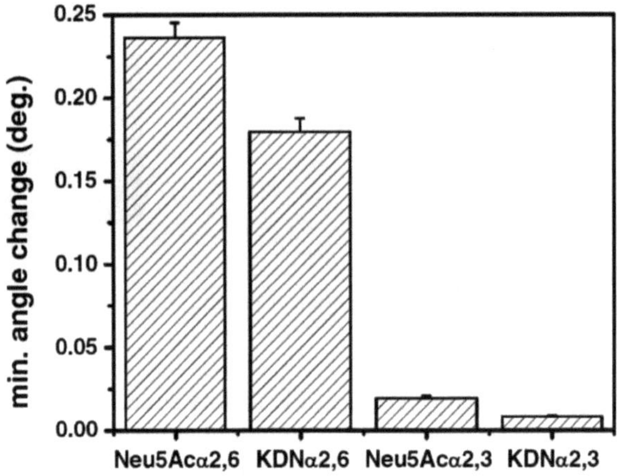

Fig. 6 Relative affinity of SNA at saturation concentration on surfaces immobilized with different biotinylated sialosides

quantitative manner. Lectin selectivity was probed by exposing one specific carbohydrate structure with known affinity to SNA to various other lectins. From these results, it is evident that this surface retains a high degree of native affinity for the carbohydrate motifs, allowing distinction of sialyl linkages, investigation of the effect of functional group on binding efficiency, and determination of rate and equilibrium constants. This surface design can be easily modified to identify and quantify the binding patterns of low-affinity biological systems, opening new avenues for probing carbohydrate–protein interactions in real time.

Recently, Vornholt et al. reported a cuvette-based SPR method for examining binding domains of lectins on carbohydrate-immobilized surfaces [19]. By taking advantage of the low sample volume required by their cuvette-based method, screening of lectin binding was performed. The assay was especially useful for more expensive lectins where sample restrictions exist.

Since protein–carbohydrate interactions are important and occur throughout cells, high-throughput screening of these interactions is urgently needed. This has been much of the focus of SPR imaging technology which will be covered later in the review. Most recently, a surface plasmon resonance-based natural carbohydrate microarray for screening of interactions between carbohydrates and carbohydrate-binding proteins (CBPs) has been reported [20]. The microarray allowed the real-time and simultaneous screening for recognition by CBPs without the need of fluorescent labeling. Specifically, the generated SPR glycan array presented a subset of the glycan repertoire of a human parasite *Schistosoma mansoni* where simultaneous detection of glycan-specific serum antibodies and the anti-glycan antibody profiles from sera of *S. mansoni*-infected were recorded. The SPR assay was sensitive to slight differences between infection sera and control sera, and revealed antibody titers and antibody classes (IgG or IgM). These results indicate that SPR-based arrays constructed from glycans can be used as unique analytical tools for screening infection markers.

3.2 Protein–DNA and Protein–Protein Interactions

Related to protein–carbohydrate interactions are protein–aptamer studies. Aptamers are short single-stranded oligonucleotide ligands chosen from large oligonucleotide libraries by an in vitro evolution process known as SELEX (systematic evolution of ligands by exponential enrichment). Aptamers bind many biomolecules, including proteins, with high specificity due to their specific base sequence and steric configuration [21]. Because aptamers are generally more stable than antibodies and are easier to generate, they have obvious use in SPR biosensor technology [22]. One group recently developed an aptamer-based surface plasmon resonance biosensor to detect retinol-binding protein 4 (RBP4) in serum samples [23]. RBP4 is a useful biomarker in the diagnosis of type 2 diabetes since its level in the serum is higher in insulin-resistant states. In their experiments, a single-stranded DNA (ssDNA) aptamer was immobilized on a gold chip and RBP4 in an artificial serum mixture was injected and detected using SPR. The results show that SPR method was more sensitive than corresponding ELISA assays and yielded better dose-dependent responses. This work is significant because it outlines an approach to possibly detect a type 2 diabetes biomarker *in vivo*. Another group recently developed a real-time qualitative assay for probing the pattern of bio-molecular interaction between the human IgE and its corresponding aptamer by SPR [21]. To amplify the SPR signals, biomacromolecules such as streptavidin and anti-hIgE antibody were utilized. Their results offer some unique information about the human IgE–aptamer complex in a nonlabeling manner and provide an important analytical strategy for a greater understanding of the aptamer–protein complex.

An SPR assay for protein–DNA interactions was recently reported for measuring biological events in its natural environment. Specifically, a two-step antibody approach was developed for the study of estrogen receptor α (ERα)–DNA interactions, in which nuclear extracts prepared from MCF-7 breast cancer cells were used as the source of ERα protein [24]. The authors indicate this two-step antibody approach could be extended to any transcriptional factors, given the availability of good quality of primary antibodies. Meanwhile, another group used SPR to monitor the binding of mammalian high-mobility-group transcriptional factor (HMGA2) to target sites on immobilized DNA, and a competition assay for inhibition of the HMGA2–DNA complex was designed [25]. HMGA2 targets the DNA minor groove and plays critical roles in disease processes from cancer to obesity. Assay results indicated that the protein binds in a 1:1 complex to two closely spaced DNA sequences that have five or six adjacent AT base pairs. The kinetics for the binding events was monitored, in the hope of laying a framework for possible elucidation of HMGA2 inhibitors.

Protein–protein interactions have been extensively probed with SPR and the readers are encouraged to consult the general review on protein–protein interactions by Berggard et al. [26]. An interesting recent example is the work by the Love group on a fusion protein known as smGN, which comprises of soluble green fluorescent protein (smGFP) and the calmodulin (CaM)-binding protein calspermin [27]. CaM is known as a Ca^{2+} sensor and is important because of its complex involvement in cell signal transduction pathways. Using SPR spectroscopy, the binding kinetics between immobilized smGN and calmodulin (CaM) was determined. Specifically, the binding strength and affinity of a newly synthesized fusion protein to CaM was quantitatively characterized, suggesting that smGN could possibly be used as a CaM inhibitor as well as provide information about the role of CaM *in vivo*.

3.3 Protein–Lipid Interactions

The newest area of SPR biological binding studies is protein–lipid interactions. These studies are important as many lipid–protein complexes modulate cellular functions and control various signal transduction cascades [28]. Ito et al. reported a kinetic study on the interactions between mannosylerythritol lipid (MEL)-assembled monolayers and various classes of immunoglobulins, including HIgG, HIgA, and HIgM using SPR [29]. The effect of MEL structure on the binding behavior of HIgG was examined, and SPR kinetic data enabled the identification of important binding motifs in this protein–lipid interaction. Another protein–lipid interaction study involves interactions between unmodified natural lipids and lipid transfer proteins (LTPs) [30]. The authors used genetically engineered biotinylated peptides to anchor LDPs on the surface of prefabricated chips to examine their interactions with various lipid molecules. Rate and equilibrium constants were determined.

We have taken a different approach to study protein–lipid interactions. As the attaching chemistry is well developed on glass substrates rather than on gold, a nanoglassified substrate has been fabricated using a conventional gold chip [31]. The fabrication process is straightforward, exhibits nanomolar sensitivity for protein analytes, and allows direct vesicle fusion on a glass-like surface, which can be conveniently monitored by SPR. We subsequently employed this nanoglassified surface to create lipid arrays to examine the cholera toxin (CT)-monosialotetrahexosylganglioside (GM1) interaction with SPR imaging [32]. More recently, a microfluidic version of a tethered bilayer membrane (tBLM) was developed on nanoglassified substrates to monitor the protein–ligand interaction in an array format [33]. In this work, the tBLM arrays demonstrate marked stability and high mobility, which provide an ideal host environment for membrane-associated proteins and open new avenues for high-throughput analysis of these proteins.

4 SPR Imaging and Microarray Technology

SPR imaging technology has attracted considerable attention recently and its application has reached a sizable scale with the advent of more available commercialized instrumentation. The label-free nature of SPRi makes it an attractive alternative to fluorescence assays in which the labeling process is complicated, can cause protein denaturation, and has inherent high background signals from intrinsic fluorescence [34]. This section reviews research development in the areas of SPRi and microarray technology in the last few years.

4.1 SPRi Examination of Protein–Protein or Protein–DNA/Aptamer Interactions

As mentioned earlier in this chapter, protein–aptamer or DNA interactions have become a quite active research area as high-throughput analysis is needed to monitor hundreds of these interactions simultaneously. We recently demonstrated a multiplexed, simultaneous analysis of antigen–antibody interactions that involve human immunoglobulin G (IgG) on a gold substrate by SPRi [35]. A multichannel, microfluidic chip was fabricated from poly(dimethylsiloxane) (PDMS) to selectively functionalize the surface and deliver the analyte solutions. Four mouse antihuman IgG antibodies were selected for evaluation, and the screening was achieved by simultaneously monitoring the protein–protein interactions under identical conditions. This assay achieved nanomolar detection sensitivity for IgG in treated serum samples. In addition, the use of PDMS multichannels for affinity studies of DNA aptamer–human Immunoglobulin E (IgE) interactions was also reported by SPRi [36]. The Zare group reported a PDMS microfluidic device containing an array of gold spots onto which antigens or antibodies of interest are attached for SPRi detection [37]. Antigen–antibody reactions were detected and quantitatively characterized in about 10 min at the subnanomolar level. To increase the sensitivity of the assay, gold nanoparticles were selectively coupled to the immunocomplex, resulting in sensitivity reaching the 10–100 pM level with limited sample volume requirements.

Ladd et al. recently demonstrated simultaneous detection of DNA and proteins in an array format with a home-built SPR imager [38]. In this work, a DNA array was created, and then part of the DNA array was converted into a protein array for simultaneous detection of two cDNA sequences and two human pregnancy hormones. Sato and colleagues reported on the mechanism of the noncross-linking gold nanoparticle (GNP) between DNA duplex-modified GNPs and a DNA duplex-modified flat gold surface with SPRi [39]. Beyond the protein–protein or protein–DNA interaction work, there have also been new reports of using SPR to detect DNA–DNA interactions [40] and viral RNA–protein interactions [41].

4.2 New Array and Signal Amplification Technology with SPRi

Currently, one of the major drawbacks of SPRi technology for high-throughput screening applications is its lack of sensitivity compared to fluorescence. Thus, many different methods have been formulated to enhance detection signals in order to compete with fluorescence methods. One approach to improve SPRi sensitivity is to develop new methods or new instrumentation to examine SPR images. Singh and Hillier [42] reported on a variant of SPRi that utilizes surface plasmon resonance dispersion as a mechanism to provide multicolor contrast for imaging thin molecular films. In this technique, colors transform in response to the formation of thin films on the surface. To demonstrate the applicability of their approach, a protein microarray was formed by a commercial ink jet printer and submonolayer films of a test protein (bovine serum albumin) were detected. Another interesting approach for SPRi is the coupling of SPRi to imaging ellipsometry (IE) to monitor the thickness of phospholipid films of a varying number of layers [13,14]. An SPRi and IE constructed with a single optical system mounted on a goniometer allowed thickness measurements to be sequentially performed on the same area of a sample. Switching between the two instrumental modes was quick and the researchers determined SPRi was better suited to measure thin films of a few nanometers while IE worked better for thicker ones. In addition, Beusink and coworkers reported monitoring the SPR image by continuous angle scanning of the SPR angle rather than the change in reflectivity at a fixed angle [43]. This resulted in a 10-fold increase in the linearity of the interactions compared with conventional SPRi fixed angle measurements and allowed for individual imaging of each array element on a sensing surface in real time.

Chemical and biological methods have also been used for signal amplification in SPRi. Corn and coworkers developed a method for the detection of protein

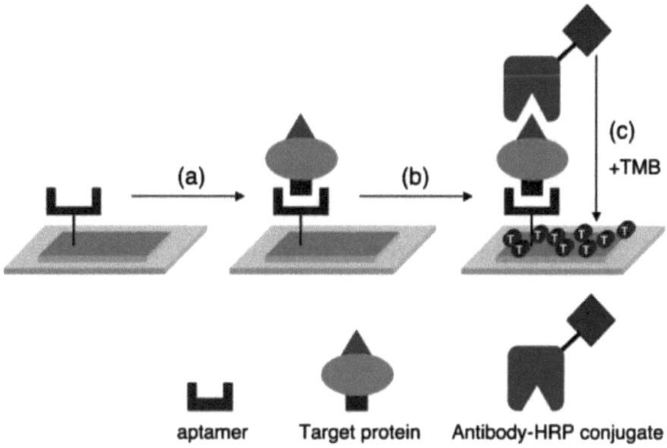

Fig. 7 A schematic illustration of the enzymatically amplified SPRi detection of target proteins (from [44])

biomarkers at picomolar concentrations that utilizes SPRi measurements of RNA aptamer microarrays [44]. In this method, the SPRi response signal is augmented using enzymatic amplification. A schematic of this process is illustrated in Fig. 7. Briefly, an RNA aptamer/protein biomarker/antibody-HRP (horseradish peroxidase) sandwich motif is formed on the microarray surface, and a subsequent localized HRP-TMB (3,3-,5,5-tetramethylbenzidine) precipitation reaction is used to amplify the SPRi response due to specific protein bio-marker adsorption onto the RNA aptamer array. This enzymatically amplified SPRi methodology led to detection of human thrombin at high fM levels and protein vascular endothelial growth factor (VEGF) at a biologically relevant concentration of 1 pM, putting SPRi in the close range of fluorescence detection. Inoue et al. reported on the enzymatic activity monitoring of caspases by using peptide arrays with SPRi [45]. Signal enhancement was achieved using streptavidin and surface-immobilized substrate peptides labeled with biotin, while the cleavage of the substrate peptides by caspases leads to a signal decrease. Using this method, they were able to monitor the activities of caspases in cell lysates, making this assay useful for drug screening purposes. Other methods to enhance analytical signals from SPRi include the use of gold nanoparticles [37] and gold nanoposts [46].

4.3 SPRi and Carbohydrate Microarrays

Examining protein–carbohydrate interactions in an array format with immobilized carbohydrates has also been reported. In addition to the three reports by our group on protein–carbohydrate interactions with SPRi [31, 32, 47], the Livache group has developed a polypyrrole-based oligosaccharide chip constructed via a copolymerization process of pyrrole and pyrrole-modified oligosaccharide to screen protein–carbohydrate interactions [48]. They covalently immobilized various carbohydrate probes and studied affinity binding patterns for different glycosaminoglycan (GAG) fragments and proteins. The detection limit was in the low nanomolar range. Another carbohydrate microarray method came out recently from a much more biological perspective. Karmanska and coworkers reported plant lectin recognition of glycans by SPR imaging using a model carbohydrate microarray based on biotin–NeutrAvidin interactions [49]. SPR imaging of an array of 40 sialylated and unsialylated glycans established the binding preferences to various carbohydrate moieties.

5 Wave of the Future: Hyphenated SPR Techniques

One of the most exciting developments with SPR is the development of coupled analytical techniques with surface plasmon resonance. While many other coupled analytical techniques such as LC-MS and GC-MS are well practiced in routine work, little research has been done coupling SPR to other analytical tools until very recently. The most promising coupling technique continues to be SPR-mass spectrometry (MS) and LC-SPR-MS. LC-SPR and SPR-fluorescence (SPFS) have gained considerable attention in recent years. We briefly review these hyphenated techniques and discuss their prospects in future SPR measurements.

5.1 SPR-MS and LC-SPR-MS

SPR is a quantitative method that generally lacks in identification capability. Coupling SPR to the most powerful identification tool, mass spectrometry, would be very analytically useful. Marchesini and coworkers developed the online nanoscale coupling of SPR for the screening of low molecular weight molecules with nanoliquid-chromatography electrospray ionization time-of-flight mass spectrometry (nano-LC ESI TOF MS) [50]. The interface is based on a chip that contains antibodies raised against the analyte thus allowing the characterization of the sensing surface with SPR. The analytical procedure has four stages: (1) sample preparation, (2) screening on chip with SPR, (3) sample capture on a separate chip, and (4) analyte desorption and analysis with nano-LC ESI TOF MS. This coupling interface enables the screening of small molecules followed by identity confirmation in suspected noncompliant samples.

Visser and coworkers combined LC, SPR, and MS together and immobilized cGMP molecules to an SPR chip to monitor the binding and dissociation of proteins from a human lysate by sequential elution steps and SPR [51]. The eluted proteins were thereafter identified by LC-MS/MS. The data indicate that SPR-based chemical proteomics is a viable alternative for quantitative extraction and identification of small-molecule-binding proteins from complex matrices. In fact, SPR-MS technological innovations have grown considerably in the last few years and have even sparked enough interest that leads to a book chapter devoted to the topic [52].

A highly interesting development is the creation of an SPR-MS array platform and realization of SPR/MS detection on a single high-content protein microarray [53, 54]. Briefly, antibodies to five human plasma proteins were arrayed on a chemically activated gold chip and binding of proteins to their corresponding antibodies was monitored via SPR imaging. Following the protein affinity retrieval, the chip was overlaid with MALDI matrix and MS analyzed, producing protein-specific mass spectra from distinct spots on the array. The SPR-MS dual detection demonstrates that both protein concentration as well as structural aspects of protein variants can be detected. High-throughput detection using various forms of SPR-MS have also been reported for automated affinity purification of recombinant and native proteins [55] and identification of proteins captured on DNA surfaces [56]. These examples aforementioned represent just the beginning in the rapidly expanding field of SPR/MS technology.

5.2 HPLC-SPR and SPFS

Recently, Zhou's group revealed a simplistic approach to couple HPLC with SPR for continuous separation and detection of protein samples [57]. The detection was realized by electrostatic interactions between the functionalized sensor surface and the charged protein analytes. This method detected changes at the sensor surface with greater sensitivity (micromolar range) than conventional methods in complex matrices, marking a huge step forward in coupling these two popular techniques together that give impressive performance.

Another interesting technique in SPR is surface plasmon fluorescence spectroscopy (SPFS), which was developed by Knoll and coworkers in 2002 [58]. This technique enables SPR detection along with enhanced fluorescence detection simultaneously. Recently, Knoll's group used SPFS to measure the thickness and photoluminescence of functional organic films on a gold substrate [59]. In addition, there is a recent report on the detection homogeneous DNA with SPFS by using labeled peptide nucleic acids (PNAs) [60]. The use of surface plasmons for exciting fluorescence enhances the signal, but the labeling process necessary for SPFS makes the measurement a bit counterintuitive as compared to label-free SPR. Nonetheless, coupling SPR with fluorescence marks a unique way to characterize biological binding in a highly integrated platform that may offer complimentary results.

6 Concluding Remarks

In the last 3 years, advances in instrumentation and surface functionalization have made SPR biosensor applications more attractive than ever for probing various biological interactions. In this chapter, we have reviewed SPR bio-interaction analysis, SPR imaging with microarrays, and various coupled analytical techniques with SPR. All of these research areas have greatly expanded the field of SPR biosensing, and there is reason to believe that much more can be accomplished in this field to provide simple, fast, sensitive, and selective assays to examine biological interactions in its native environment and in real time. With no labeling process necessary for SPR bio-interaction analysis, biocompatibility issues are generally avoided and the requirement of sophisticated optical systems is reduced, making this technique immensely popular in the biology community. While this review has focused on innovation with regards to biointerfaces and SPR instrumentation, many routine analyses can be performed with SPR as evidenced by the abundance of literature. Readers are encouraged to refer to the reviews that focus on the applications and analysis using standard chips. Highly innovative research in surface chemistry and SPR methodologies continues to thrive as they provide edging technical development. With the arrival of inexpensive instrumentation, SPR is becoming more readily available to more groups specializing in surface sciences. In addition, new chips are being commercialized by companies like Biacore and Eco Chemie allowing the routine user to take advantage of the innovation. As the multitude of groups using SPR for bio-sensing applications grows, many new areas will emerge. End users such as pharmaceutical industry, homeland defense, and medical diagnostic services will greatly benefit from the multidisciplinary nature of SPR innovation and usage for many years to come.

References

1. Homola J (2008) Surface plasmon resonance sensors for detection of chemical and biological species. Chem Rev 108:462–493
2. Liedberg B, Lundstrom I, Stenberg E (1993) Principles of biosensing with an extended coupling matrix and surface plasmon resonance. Sensors Actuators B Chem 11:63–72
3. Smith EA, Corn RM (2003) Surface plasmon resonance imaging as a tool to monitor biomolecular interactions in an array based format. Appl Spectrosc 57:320A–332A
4. Knoll W (1998) Interfaces and thin films as seen by bound electromagnetic waves. Annu Rev Phys Chem 49:569–638
5. Raether H (1988) Surface plasmons on smooth and rough surfaces and on gratings springer tracts in modern physics. Springer, Berlin
6. Wiltschi B, Knoll W, Sinner E-K (2006) Binding assays with artificial tethered membranes using surface plasmon resonance. Methods 39:134–146
7. Frutos AG, Corn RM (1998) SPR of ultrathin organic films. Anal Chem 70:449A–455A
8. Hashimoto S, Isobe T, Natsume T (2007) Biomolecular interaction analysis coupled with mass spectrometry to detect interacting proteins. In: Walker JM (ed) The proteomics protocols handbook. Humana Press, Totowa, NJ
9. Redman JE (2007) Surface plasmon resonance for probing quadruplex folding and interactions with proteins and small molecules. Methods 43:302–312
10. Jönsson U, Malmqvist M (1992) Real-time biospecific interaction analysis. Adv Biosensors 2:291–336
11. Edwards PR, Leatherbarrow RJ (1997) Determination of association rate constants by an optical biosensor using initial rate analysis. Anal Biochem 246:1–6

12. Navratilova I, Myszka DG (2006) Investigating biomolecular interactions and binding properties using spr biosensors. Surface plasmon resonance based sensors. Springer, Berlin, pp 159–161
13. Li B, Che J, Long M (2008) Measuring binding kinetics of surface-bound molecules using the surface plasmon resonance technique. Anal Biochem 377:195–201
14. Li Y-J, Zhang Y, Zhou F (2008) Sequential monitoring of film thickness variations with surface plasmon resonance imaging and imaging ellipsometry constructed with a single optical system. Anal Chem 80:891–897
15. Steiner G (2004) Surface plasmon resonance imaging. Anal Bioanal Chem 379:328–331
16. Hakomori S (2004) Carbohydrate-to-carbohydrate interaction, through glycosynapse, as a basis of cell recognition and membrane organization. Glyconjugate J 21:125–137
17. Monsigny M, Mayer R, Roche AC (2000) Sugar–lectin interactions: sugar clusters, lectin multivalency and avidity. Carbohydr Lett 4:35–52
18. Linman MJ, Taylor JD, Yu H, Chen X, Cheng Q (2008) Surface plasmon resonance study of protein – carbohydrate interactions using biotinylated sialosides. Anal Chem 80:4007–4013
19. Vornholt W, Hartmann M, Keusgen M (2007) SPR studies of carbohydrate–lectin interactions as useful tool for screening on lectin sources. Biosens Bioelectron 22:2983–2988
20. de Boer AR, Hokke CH, Deelder AM, Wuhrer M (2008) Serum antibody screening by surface plasmon resonance using a natural glycan microarray. Glycoconj J 25:75–84
21. Wang J, Lv R, Xu J, Xu D, Chen H (2008) Characterizing the interaction between aptamers and human IgE by use of surface plasmon resonance. Anal Bioanal Chem 390:1059–1065
22. Lee JF, Stovall GM, Ellington AD (2006) Aptamer therapeutics advance. Curr Opin Chem Biol 10:282–289
23. Lee SJ, You B-S, Park JW, Niazi JH, Kim YS, Gu MB (2008) ssDNA aptamer-based surface plasmon resonance biosensor for the detection of retinol binding protein 4 for the early diagnosis of type 2 diabetes. Anal Chem 80:2867–2873
24. Su X, Neo SJ, Pek W, Thomsen JS (2008) A two-step antibody strategy for surface plasmon resonance spectroscopy detection of protein–DNA interactions in nuclear extracts. Anal Biochem 376:137–143
25. Miao Y, Cui T, Leng F, Wilson WD (2008) Inhibition of high-mobility-group A2 protein binding to DNA by netropsin: a biosensor-surface plasmon resonance assay. Anal Biochem 374:7–15
26. Berggård T, Linse S, James P (2007) Methods for the detection and analysis of protein-protein interactions. Proteomics 7:2833–2842
27. Murphy AJ, Kemp F, Love J (2008) Surface plasmon resonance characterization of calspermin–calmodulin binding kinetics. Anal Biochem 376:61–72
28. Feng L, Ferguson C, Nielsen PO, Chakravarty L, Rzepecki PW, Prestwich GD (2006) Methods of probing phosphoinositides-protein interactions. In: Feng LP, Prestwich GD (eds) Functional lipidomics. Taylor & Francis, Boca Raton, FL, pp 215–274
29. Ito S, Imura T, Fukuoka T, Morita T, Sakai H, Abe M, Kitamoto D (2007) Kinetic studies on the interactions between glycolipid biosurfactant-assembled monolayers and various classes of immunoglobulins using surface plasmon resonance. Colloids Surf Biointerfaces 58:165–171
30. Kernstock RM, Girotti AW (2007) Lipid transfer protein binding of unmodified natural lipids as assessed by surface plasmon resonance methodology. Anal Biochem 365:111–121
31. Phillips KS, Han J-H, Martinez M, Wang Z, Carter D, Cheng Q (2006) Nanoscale classifcation of gold substrates for surface plasmon resonance analysis of protein toxins with supported lipid membranes. Anal Chem 78:596–603
32. Phillips KS, Wilkop T, Han J-H, Wu J-J, Al-Kaysi RO, Cheng Q (2006) Surface plasmon resonance imaging analysis of protein-receptor binding in supported membrane arrays on gold substrates with calcinated silicate films. J Am Chem Soc 128:9590–9591
33. Taylor JD, Phillips KS, Cheng Q (2007) Microfluidic fabrication of addressable tethered lipid bilayer arrays and optimization using SPR with silane-derivatized nanoglassy substrates. Lab Chip 7:927–930

34. Verducci JS, Melfi VF, Lin S, Wang Z, Roy S, Sen CK (2006) Microarray analysis of gene expression: considerations in data mining and statistical treatment. Physiol Genomics 25:355–363
35. Dong Y, Wilkop T, Xu D, Wang Z, Cheng Q (2008) Microchannel chips for the multiplexed analysis of human immunoglobulin G–antibody interactions by surface plasmon resonance imaging. Anal Bioanal Chem 390:1575–1583
36. Wang Z, Wilkop T, Xu D, Dong Y, Ma G, Cheng Q (2007) Surface plasmon resonance imaging for affinity analysis of aptamer–protein interactions with PDMS microfluidic chips. Anal Bioanal Chem 389:819–825
37. Luo Y, Yu F, Zare RN (2008) Microfluidic device for immunoassays based on surface plasmon resonance imaging. Lab Chip 8:694–700
38. Ladd J, Taylor AD, Pilarik M, Homola J, Jiang S (2008) Hybrid surface platform for the simultaneous detection of proteins and DNAs using a surface plasmon resonance imaging sensor. Anal Chem 80:4231–4236
39. Sato Y, Hosokawa K, Maeda M (2008) Detection of non-cross-linking interaction between DNA-modified gold nanoparticles and a DNA-modified flat gold surface using surface plasmon resonance imaging on a microchip. Colloids Surf Biointerfaces 62:71–76
40. Manera MG, Spadavecchia J, Leone A, Quaranta F, Rella R, Dell'atti D, Minunni M, Mascini M, Siciliano P (2008) Surface plasmon resonance imaging technique for nucleic acid detection. Sensors Actuators B Chem 130:82–87
41. Garcia BH II, Goodman RM (2008) Use of surface plasmon resonance imaging to study viral rna: protein interactions. J Virol Methods 147:18–25
42. Singh BK, Hillier AC (2007) Multicolor surface plasmon resonance imaging of ink jet-printed protein microarrays. Anal Chem 79:5124–5132
43. Beusink JB, Lokate AMC, Besselink GAJ, Pruijn GJM, Schasfoort RBM (2008) Angle-scanning SPR imaging for detection of biomolecular interactions on microarrays. Biosens Bioelectron 23:839–844
44. Li Y, Lee HJ, Corn RM (2007) Detection of protein biomarkers using RNA aptamer microarrays and enzymatically amplified surface plasmon resonance imaging. Anal Chem 79:1082–1088
45. Inoue Y, Mori T, Yamanouchi G, Han X, Sonoda T, Niidome T, Katayama Y (2008) Surface plasmon resonance imaging measurements of caspase reactions on peptide microarrays. Anal Biochem 375:147–149
46. Malic L, Cui B, Veres T, Tabrizian M (2007) Enhanced surface plasmon resonance imaging detection of DNA hybridization on periodic gold nanoposts. Opt Lett 32:3092–3094
47. Linman MJ, Yu H, Chen X, Cheng Q (2009) Fabrication and characterization of a sialoside-based carbohydrate microarray biointerface for protein binding analysis with surface plasmon resonance imaging. ACS Appl Mater Interfaces 1:1755–1762
48. Mercey E, Sadir R, Maillart E, Roget A, Baleux F, Lortat-Jacob H, Livache T (2008) Polypyrrole oligosaccharide array and surface plasmon resonance imaging for the measurement of glycosaminoglycan binding interactions. Anal Chem 80:3476–3482
49. Karamanska R, Clarke J, Blixt O, MacRae JI, Zhang JQ, Crocker PR, Laurent N, Wright A, Flitsch SL, Russell DA, Field RA (2008) Surface plasmon resonance imaging for real-time, label-free analysis of protein interactions with carbohydrate microarrays. Glycoconj J 25:69–74
50. Marchesini GR, Buijs J, Haasnoot W, Hooijerink D, Jansson O, Nielen MWF (2008) Nanoscale affinity chip interface for coupling inhibition SPR immunosensor screening with nano-LC TOF MS. Anal Chem 80:1159–1168
51. Visser NFC, Scholten A, van den Heuvel RHH, Heck AJR (2007) Surface-plasmon-resonance-based chemical proteomics: efficient specific extraction and semiquantitative identification of cyclic nucleotide-binding proteins from cellular lysates by using a combination of surface plasmon resonance, sequential elution and liquid chromatography–tandem mass spectrometry. Chem Bio Chem 7:298–305

52. Nedelkov D, Nelson RW (2006) Surface plasmon resonance mass spectrometry for protein analysis. In: Nedelkov DN, Nelson RW (eds) Methods in molecular biology: new and emerging proteomic techniques. Humana Press, Totowa, NJ, pp 131–139
53. Nedelkov D, Tubb KA, Nelson RW (2006) Surface plasmon resonance-enabled mass spectrometry arrays. Electrophoresis 27:3671–3675
54. Nedelkov D (2007) Development of surface plasmon resonance mass spectrometry array platform. Anal Chem 79:5987–5990
55. Borch J, Roepstorff P (2006) Combinations of SPR and MS for characterization of native and recombinant proteins in cell lysates. Mol Biotechnol 33:179–190
56. Bouffartigues E, Leh H, Anger-Leroy M, Rimsky S, Buckle M (2007) Rapid coupling of surface plasmon resonance (SPR and SPRi) and proteinchip based mass spectrometry for the identification of proteins in nucleoprotein interactions. Nucl Acids Res 35:e39
57. Du M, Zhou F (2008) Postcolumn renewal of sensor surfaces for high-performance liquid chromatography – surface plasmon resonance detection. Anal Chem 80:4225–4230
58. Neumann T, Johansson M-L, Kambhampati D, Knoll W (2002) Surface-plasmon fluorescence spectroscopy. Adv Funct Mater 12:575–586
59. Mitamura K, Imae T, Tian S, Knoll W (2008) Surface plasmon fluorescence investigation of energy-transfer-controllable organic thin films. Langmuir 24:2266–2270
60. Chu L-Q, Forch R, Knoll W (2007) Surface-plasmon-enhanced fluorescence spectroscopy for DNA detection using fluorescently labeled PNA as "DNA Indicator". Angew Chem Int Ed 46:4944–4947

Nanohole Arrays in Metal Films as Integrated Chemical Sensors and Biosensors

Alexandre G. Brolo, Reuven Gordon, and David Sinton

Abstract Ordered arrays of subwavelength holes in optically thick metal films exhibit optical properties that may be exploited to achieve chemical and biological sensing. The fundamental phenomena governing these interactions, the sensing methodologies they enable, and the on-chip integration of nanohole array sensors are described in this chapter. The fundamental phenomena of confinement, or guiding of electromagnetic waves at a metal surface that are central to the sensing capabilities offered by nanohole arrays in metal films are described first. The fundamental basis for surface plasmon resonance on smooth planar metal-dielectric interfaces as well as the extension and localization of these phenomena to nanostructures is described. Nanohole-array-based sensing methodologies are discussed next. The extraordinary optical transmission through nanohole arrays is described with the application of that phenomenon to surface plasmon resonance-based sensing. Field localization, related to the surface plasmon excitation, enables surface-enhanced Raman scattering (SERS) and surface-enhanced fluorescence spectroscopy (SEFS). The application of nanohole arrays in these sensing methodologies are described, as are recent efforts to further localize the electromagnetic field via overlapped double-hole structures. A selection of recently presented

D. Sinton (✉)
Department of Mechanical Engineering, University of Victoria, P.O. Box 3055, STN CSC, Victoria, BC, Canada, V8W 3P6
e-mail: dsinton@me.uvic.ca

A.G. Brolo
Department of Chemistry, University of Victoria, P.O. Box 3055, STN CSC, Victoria, BC, Canada V8W 3P6,

R. Gordon
Department of Electrical and Computer Engineering, University of Victoria, P.O. Box 3055, STN CSC, Victoria, BC, Canada V8W 3P6,

M. Zourob and A. Lakhtakia (eds.), *Optical Guided-wave Chemical and Biosensors I*, 155
Springer Series on Chemical Sensors and Biosensors 7,
DOI 10.1007/978-3-540-88242-8_6,

experimental results are highlighted throughout the chapter to demonstrate the relevant phenomena and sensing capabilities. In addition to the variety of sensing opportunities offered, both the small footprint of nanohole arrays and the simplified transmission mode operation at normal incidence are highly advantageous with respect to device-level miniaturization. Finally, the micro- and nanofluidic integration of nanohole-array-based sensors is discussed. Integration efforts to date, as well as future prospects for nanohole arrays in a lab-on-chip format and potential to exploit transport phenomena in these structures to the benefit of chemical and biological sensing applications, are described.

Keywords Nanohole array · Surface plasmon resonance · Optical sensing · Chemical sensing · Biosensing · Microfluidic · Nanofluidic · Extraordinary optical transmission

Contents

Abbrreviations

ATR	Attenuated total-internal reflection
BSA	Bovine serum albumin
EOT	Extraordinary optical transmission
FDTD	Finite-difference time-domain
FIB	Focused ion beam
LSP	Localized surface plasmon
MIM	Metal-insulator-metal
MUA	Mercaptoundecanoic acid
RIU	Refractive index unit
RR	Resonance Raman
SEFS	Surface enhanced fluorescence microscopy
SERS	Surface enhanced Raman scattering
SERRS	Surface-enhanced resonance Raman scattering
SHG	Second harmonic generation

SP	Surface plasmon
SPP	Surface plasmons polaritons
SPR	Surface plasmon resonance
TM	Transverse magnetic

Symbols

$\varepsilon_{d,m}$	Relative permittivity of dielectric, metal
ε_0	Permittivity of vacuum
ε_r	Relative permittivity
ω	Angular frequency of light
ω_p	Angular plasma frequency
c	Speed of light in vacuum
d	Diameter
D	Molecular diffusivity
$\bar{E}$	Electric field
$\bar{H}$	Magnetic field
k	Reaction rate constant
n	Refractive index
n_{eff}	Effective refractive index
p	Periodicity
T	Transmittance
x, y, z	Coordinate directions

1 Introduction to Optical Sensing Using Metals

The confinement, or guiding, of electromagnetic waves at a metal surface is central to the chemical and biological sensing capabilities offered by nanohole arrays in metal films. In this section, the fundamental phenomena governing these interactions, as well as the extension of these phenomena to nanostructures are described. This section provides a theoretical background for the sensing methodologies described in this chapter; a more comprehensive coverage of plasmonics fundamentals and applications is provided by Maier [1].

1.1 An Introduction to Surface Plasmon Polaritons

Surface plasmons polaritons (SPPs) are classically described as electromagnetic waves formed by charge oscillations at the surface of a metal. Figure 1a shows

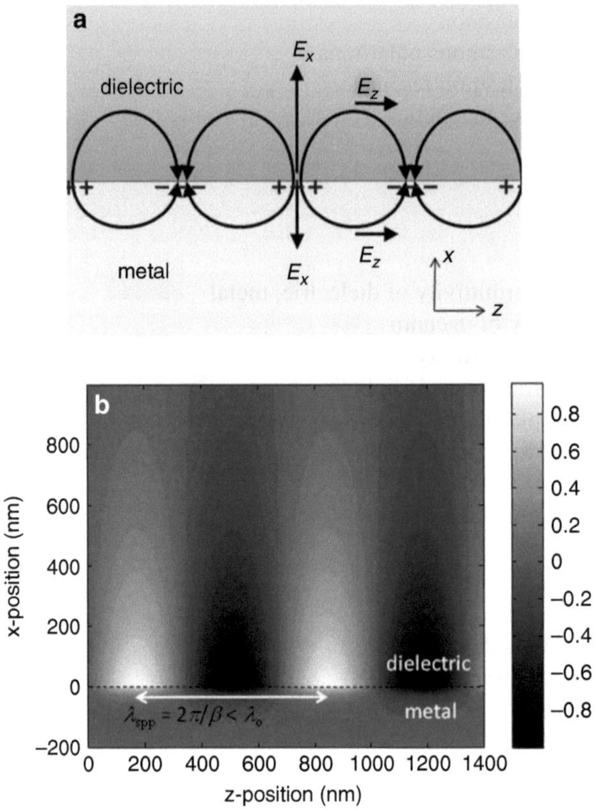

Fig. 1 (**a**) Schematic representation of SPP as charge oscillations at the interface between a metal and a dielectric. It is clear from this picture that the electric field has a longitudinal (z-direction) component that is $\pi/2$ out of phase with the transverse component (x-direction). (**b**) Calculated transverse magnetic field (y-direction) for an SPP above gold at free-space wavelength of 700 nm. The SPP wavelength is shorter than the free-space wavelength, as described in the text

a commonly used picture, which is useful for visualizing the SPP. From this figure, we see that the electrons in the metal have moved to create positive and negative charges, which create electric polarization. The electric field normal to the surface is out-of-phase by $\pi/2$ with respect to the electric field parallel to the surface. The wave has a transverse magnetic component, in the y-direction.

Solving Maxwell's equations with exponential solutions decaying away from the interface and propagation in the plane of the interface gives the form of the y-component of the SPP magnetic field as

$$H_y(x,z,t) = \begin{cases} \exp(-\gamma_d x + i\beta z - i\omega t), & x > 0 \\ \exp(\gamma_m x + i\beta z - i\omega t), & x < 0 \end{cases}, \tag{1}$$

with

$$\gamma_{\mathrm{m,d}} = \frac{\omega}{c}\sqrt{\frac{-{\varepsilon_{\mathrm{m,d}}}^2}{\varepsilon_{\mathrm{m}} + \varepsilon_{\mathrm{d}}}} \tag{2}$$

and

$$\beta = \frac{\omega}{c}\sqrt{\frac{\varepsilon_{\mathrm{m}}\varepsilon_{\mathrm{d}}}{\varepsilon_{\mathrm{m}} + \varepsilon_{\mathrm{d}}}}, \tag{3}$$

where $i = \sqrt{-1}$, $\omega = 2\pi f$ is the angular frequency of light; c is the speed of light in vacuum; ε_{m} is the relative permittivity of the metal; and ε_{d} is the relative permittivity of the dielectric. Equation (1) describes a waveguide mode when $\varepsilon_{\mathrm{m}} \leq \varepsilon_{\mathrm{d}}$. The sign change in the relative permittivity occurs in the optical regime for some metals, and this causes a phase-shift of π in the x-component of the electric field when crossing the boundary. Crossing the boundary twice gives the 2π phase-shift required for self-consistency of the waveguide mode. Therefore, the negative value of the relative permittivity of the metal is essential to waveguiding of the SPP mode. Here, we have ignored the typically smaller imaginary contributions that lead to loss. Loss can be captured by the imaginary part of β, the propagation constant in (3). Figure 1b shows the transverse magnetic field of the SPP wave, which is a TM waveguide mode.

The remaining electric field components can be found by using the Ampère-Maxwell equation:

$$\nabla \times \vec{H} = -i\omega\varepsilon_0\varepsilon_{\mathrm{r}}(\omega, z)\vec{E}\,, \tag{4}$$

where

$$\varepsilon_{\mathrm{r}}(\omega, z) = \begin{cases} \varepsilon_{\mathrm{d}}(\omega), & x > 0 \\ \varepsilon_{\mathrm{m}}(\omega), & x < 0 \end{cases}. \tag{5}$$

The Drude free-electron theory can be used to model the optical properties of noble metals. This theory provides the following expression for relative permittivity as a function of frequency in the visible regime (with electron scattering neglected because the electromagnetic frequency is high):

$$\varepsilon_{\mathrm{m}}(\omega) = 1 - \frac{{\omega_{\mathrm{p}}}^2}{\omega^2}, \tag{6}$$

where ω_{p} is the angular plasma frequency. This shows how the real part of the relative permittivity of a metal can be negative for $\omega < \omega_{\mathrm{p}}$. Considering (3) and

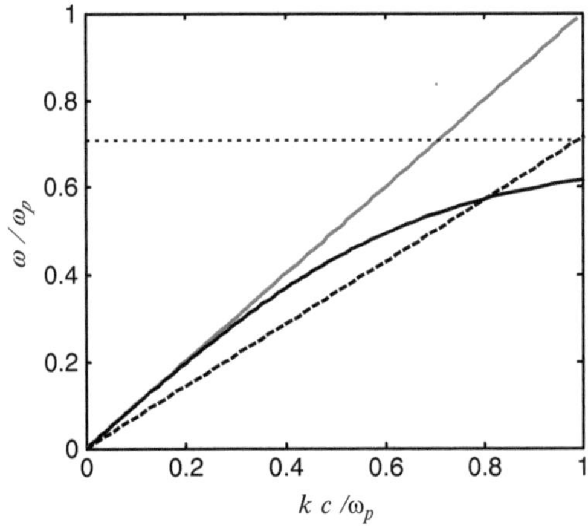

Fig. 2 Dispersion relation for SPP (shown with *solid black line*) with asymptote at $\omega \rightarrow \omega_p/2$. As the dispersion of plane wave in dielectric (*solid gray line*) does not cross SPP dispersion, phase-matching is not possible – the SPP is a waveguide mode that does not couple to radiation. Dispersion of plane wave in a higher refractive index material (*dashed black line*) is used to obtain phase-matching and efficient coupling is obtained at the intersection point with the SPP dispersion

(6), for the case where the top dielectric has $\varepsilon_d = 1$, we see that β diverges when $\omega \rightarrow {}^{\omega_p}/_2$, leading to very short wavelengths. The effective index of the SPP is:

$$n_{\text{eff}} = c\beta/\omega \tag{7}$$

which is greater than the refractive index of the dielectric, considering (3). In other words, the SPP has a shorter wavelength than plane waves in the dielectric. This means that the SPP waveguide mode is not coupled to free-space plane waves. We can add a prism with a higher refractive index than the dielectric medium supporting the SPP, or a grating to couple plane waves to the SPP, both by a process of phase-matching. Figure 2 shows the dispersion relation for the SPP and the process of attenuated total-internal reflection (ATR) phase matching.

1.2 *Surface Plasmon Resonance Sensing*

SPPs are evanescently bound to the metal surface, and they are sensitive to dielectric perturbations at the surface. Consequently, metals can be used to sense the binding of molecules to the surface by a technique called surface plasmon resonance (SPR) sensing. The Kretschmann geometry, depicted in Fig. 3a, is the most common method for SPR. It uses phase-matched evanescent coupling between propagating light in a high refractive index medium and the SPP. The phase-matching is controlled by tuning the angle and/or wavelength of the incident beam and occurs for the angle where

$$n_{\text{prism}} \sin(\theta) = n_{\text{eff}}. \tag{8}$$

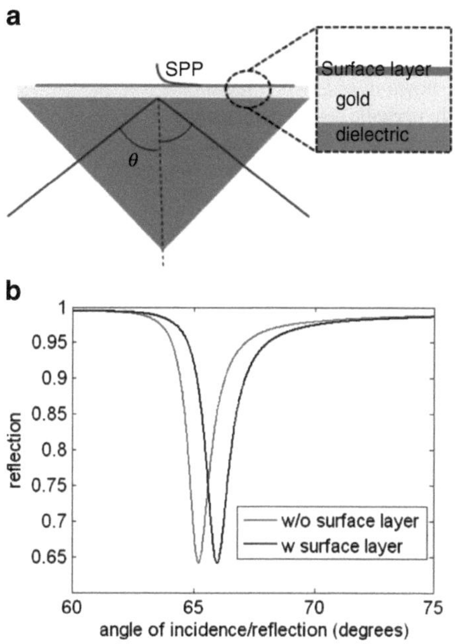

Fig. 3 (**a**) Schematic of ATR configuration for SPR sensing. The ATR configuration couples plane-waves inside the prism into SPP waves when the angle and wavelength are tuned to the phase-matching condition. Since the SPP has loss, the reflection is reduced when there is coupling to the SPP wave. (**b**) Calculated SPR reflection for a 50 nm gold layer at free-space wavelength of 632 nm, with and without a 5 nm surface layer of refractive index 1.5 (It is assumed that above the surface layer is water, with refractive index 1.33). The shift in the reflection dip is used to detect the surface layer

More precise calculations match the field components at the boundaries to find the reflection amplitude. For the phase-matched condition, light is transferred to the SPP, and so there is increased loss (from the metal) and reduced reflection.

Figure 3a shows the geometry of an ATR sensor in the Kretschmann configuration. Figure 3b shows the reflection as a function of the incidence angle when the refractive index of a 5 nm surface layer is modified to 1.5 (from 1.33 of the aqueous surrounding medium). The shift in the reflection minimum (in this case 0.7°) is used to sense the absorption of molecules at the surface. It is important to include the loss of the metal in this calculation, otherwise all the light that is coupled into the SPP is coupled out of it once again, and so the reflection is always 100%.

SPR has applications in drug development, medical diagnostics, environmental monitoring, and food and water safety [2]. Typically, the ATR SPR sensitivity is ~10^{-7} refractive index units (RIU) (i.e., the 7th significant figure after the decimal in refractive index), and the limit of detection is ~10 pg/mL.

1.3 Surface Plasmons in Nano-structured Metals

To understand the profound effect that nanostructuring can have on light, with the help of surface plasmons, we consider the simplest nanostructure: a gap between two metals.

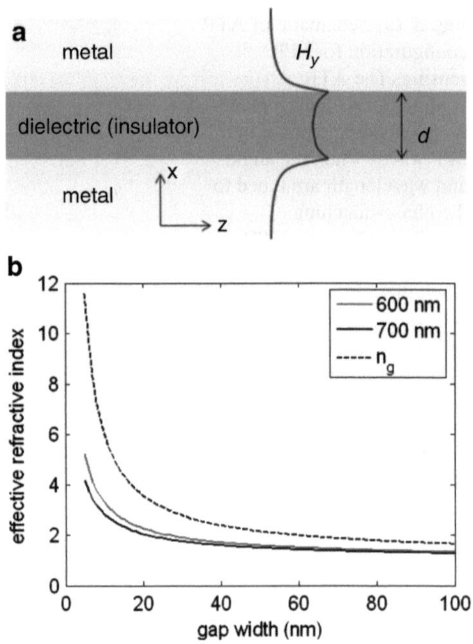

Fig. 4 (**a**) Schematic of MIM structure with gap *d*. Transverse magnetic field is shown in *red*, which has a hyperbolic cosine dependence in the gap and exponential decay into the metal. (**b**) Effective refractive index of waveguide mode within MIM gap, which increases as the gap is made smaller. Calculations for two free-space wavelengths are shown using the relative permittivity of gold and the difference is used to estimate the group index (n_g), which also increases as the gap is made smaller. Therefore, the light slows down as the gap is made smaller

Figure 4(a) shows a metal–insulator–metal (MIM) gap. The solution to the gap is equivalent to that for a TM dielectric waveguide, except that the relative permittivity of the cladding layers is negative and the field is a hyperbolic cosine inside the gap. The equation to find the solution for the effective index of the MIM waveguide mode is as follows:

$$\tanh\left(\frac{d}{2}\sqrt{\beta^2-\left(\frac{\omega}{c}\right)^2\varepsilon_\mathrm{d}}\right)=-\frac{\varepsilon_\mathrm{d}}{\varepsilon_\mathrm{m}}\sqrt{\frac{\beta^2-\left(\frac{\omega}{c}\right)^2\varepsilon_\mathrm{m}}{\beta^2-\left(\frac{\omega}{c}\right)^2\varepsilon_\mathrm{d}}}. \tag{9}$$

Figure 4b plots the effective index, $n_\mathrm{eff}=c\beta/\omega$, of the MIM waveguide mode as a function of gap width, d, for two different wavelengths. Two important results can be seen from this figure:

1. As d is made small, the effective index becomes large. In other words, the phase velocity decreases and the SPP wavelength becomes smaller: we can shrink the optical wavelength by making the gap smaller. This has many interesting effects. For example, if you want to increase the wavelength of cut-off for a rectangular hole in a metal, it is better to make the hole smaller in one direction, instead of bigger.
2. As $d$is made smaller, the group index, $\left(\frac{\partial\beta}{\partial\omega}\right)^{-1}$, increases; the SPP group velocity decreases as we make the MIM gap narrower. Therefore, if we consider a gradual taper where the gap is made narrower, light will slow down as it

progresses, and so the energy will build up (provided that loss is sufficiently small and the taper is adiabatic to minimize reflection). Strong field enhancement due to slowing-down of the surface plasmon has been demonstrated at a metal tip where the dielectric surrounds the metal.

From this, we see the two important advantages of metal nanostructures:

1. Shrinking the optical wavelength
2. Enhancing the local field intensity

Metal nanostructures (such as particles and apertures) can permit local resonances in the optical properties. These local resonances are referred to as localized surface plasmons (LSPs). The simplest version of the LSP resonance comes for a spherical nanoparticle, where the electromagnetic phase-retardation can be neglected in the quasi-static approximation, so that the electric field inside the particle is uniform and given by the usual electrostatic solution [3]:

$$E_{\mathrm{in}} = \frac{3\varepsilon_{\mathrm{d}} E_0}{\varepsilon_{\mathrm{m}} + 2\varepsilon_{\mathrm{d}}}, \tag{10}$$

where E_0 is the electric field that excites the sphere. Upon ignoring the imaginary contributions to the relative permittivities in (10), it is clear that the field inside the sphere diverges when $\varepsilon_{\mathrm{m}} = -2\varepsilon_{\mathrm{d}}$, which is the region of the LSP resonance. By virtue of the continuity of the normal component of the electric displacement vector, the electric field is enhanced by the ratio $|\varepsilon_{\mathrm{m}}/\varepsilon_{\mathrm{d}}|$ outside the sphere when the field is normal to the surface.

Clearly, the LSP resonance is also sensitive to the local refractive index (relative permittivity), and LSP have been studied for applications to SPR sensing as well [4]. One advantage of LSP is that the sensing volume is reduced to the local environment and so a smaller limit of detection and greater multiplexing are possible.

1.4 Nano-structured Metals Enhanced Optical Interactions with Materials

1.4.1 Surface-Enhanced Raman Scattering

In Raman scattering, a photon is scattered by a material with increased (or decreased) energy due to the simultaneous absorption (or emission) of a phonon (vibration mode) in the material. Raman scattering provides the phonon-energy spectrum of the material, and it is a useful spectroscopic method for identifying materials by their vibrational “fingerprint”. Raman scattering is weak – typically 14 orders of magnitude weaker than fluorescence. Therefore, we are interested in enhancing Raman scattering to routinely characterize materials down to the single-molecule level.

Surface-enhanced Raman Scattering (SERS) makes use of local field enhancements provided by metal nanostructures. The metal nanostructures focus light down to the nanoscale and enhance the local field. While this enhancement alone would usually only provide a linear increase in the Raman scattering intensity, the metal nanostructure also works in reverse to efficiently capture the Raman-scattered photons and direct them to the detector. This double-enhancement can occur because the Raman scattered photon has almost the same energy as the incident photon – the phonon energy is relatively small. The double-enhancement increases the Raman-scattering intensity by the square of the near-field intensity enhancement, even though Raman scattering is linear (in photon number). In terms of the electric field, a metal nanostructure that provides 3,000 times enhancement gives the required 14 orders of magnitude increase in Raman scattering to obtain signal at the single-molecule level.

Early observations of SERS at the single molecule level used randomly nanostructured metal surfaces [5,6].There is great potential for tailor-made nanostructures to provide more predictable field enhancements. With tailor-made nanostructures, it is possible to design the location of field enhancement, the magnitude of field enhancement, and the wavelengths that provide field enhancement.

1.4.2 Surface-Enhanced Fluorescence

As with Raman scattering, the presence of a nano-structured metal can modify the local electromagnetic environment around a molecule. This modification is commonly expressed by the Purcell factor, and it essentially means that the optical interaction with the material is enhanced by some factor. As a result, fluorescence (and other optical processes, such as absorption) can be enhanced by metal nanostructures. The disadvantage is that often the light can be strongly coupled to surface plasmons that it absorbed before it is extracted (since metals are lossy). Enhanced absorption using nano-structured metals is presently of interest for harvesting solar energy more efficiently [7].

1.4.3 Nonlinear Optics

As optical nonlinearity is typically weak and it scales nonlinearly with the electromagnetic field, there is a clear benefit from the extreme subwavelength focusing [1,8] allowed by surface plasmon nanophotonics. Strong nonlinear enhancements are possible in metal composite materials [9], or on nano-structured metal surfaces [10]. Other opportunities exist, such as using SPP/LSP resonances together to enhance different wavelengths simultaneously in frequency conversion nonlinear optics. For example, a recent work showed enhanced second harmonic generation of nanohole arrays in a metal film by matching the fundamental wavelength to the

SPP resonance and the second harmonic (at half the free-space wavelength of the fundamental) to the LSP resonance [11].

2 Nanohole-Based Sensing

2.1 Introduction to Extraordinary Optical Transmission

In 1998, Ebbesen et al. were surprised by the large amount of light transmitted through nanoholes on gold and silver films at certain wavelengths [12]. The metal films were optically thick and the hole diameters were smaller than 200 nm. The idea that the amount of light transmitted through a subwavelength aperture in an opaque metal sheet should decrease with the hole diameter was established by Bethe [13]. In fact, Bethe showed that the transmittance (T) through a single hole in an infinitely thin slab is related to the hole diameter (d) and the wavelength of the transmitted light (λ) according to $T \propto (d/\lambda)^4$.

Another well-established concept is that propagating modes can no longer be sustained inside a cylindrical hole when the hole diameter becomes smaller than about half of the wavelength of the incident light (cut-off condition). Ebbesen's experiment showed larger transmission at wavelengths beyond cut-off. Another striking observation was that there was more light transmitted than the actual amount that impinged on the hole area; i.e., the material seemed much more transparent than it should be. This phenomenon reported by Ebbesen was called "extraordinary optical transmission" (EOT). The discovery of the EOT renewed the interest on the understanding of the light transmission through apertures, and Bethe's theory was subsequently extended to single holes in real metals [14] with finite thickness [15], which confirms an exponential dependence of the transmission efficiency with the film thickness [15], and to arrays of holes in a perfect conductor [16,17].

The fact that the EOT was observed for gold and silver films, but its magnitude decreased when different metals were used hinted that SPPs might be causing the effect. Further evidence supporting the role of SPP on the EOT was the relationship between the peaks of maximum transmission and the distance between the nanoholes (periodicity), including a good fit between the experimentally obtained dispersion and the one calculated for SPP, generated by grating coupling [12]. In this case, the grating periodicity provided the extra parallel momentum required to allow matching with the free photon in units of $2\pi/p$ (p is the grating periodicity) in either the x or the y direction (defined in Fig. 5). Although the role of SPPs as the primary cause for EOT has been challenged [18,19], the experimental evidences from several laboratories seem to confirm their contribution to the effect [20–24].

The arrays of nanoholes on Au films are generally fabricated using focused ion beam (FIB) milling. The geometric parameters of the arrays, such as hole diameter and periodicity (distance between the holes), can be controlled with nanometric

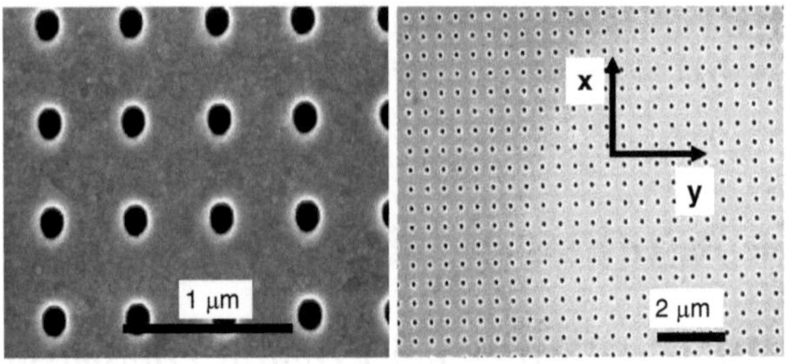

Fig. 5 Scanning electron micrographs of an array of nanoholes on gold films in two magnifications. Two hundred nanometer hole diameter and 550 nm of periodicity

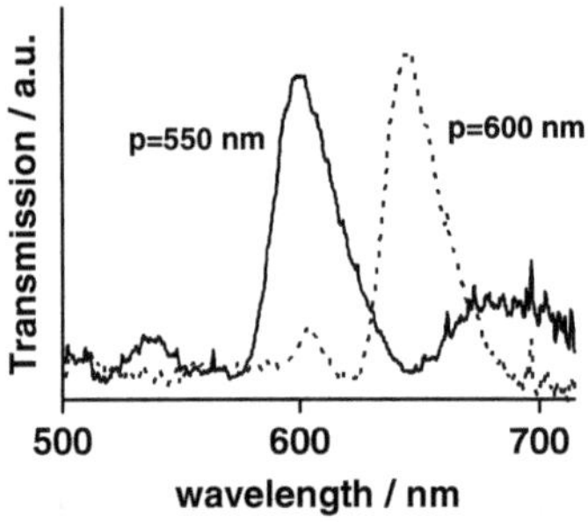

Fig. 6 EOT from two arrays of nanoholes with different periodicities (indicated). Hole diameter is 200 nm

precision using the FIB method. Figure 5 shows a scanning electron micrograph of one of the arrays used by our group. The arrays are milled in a 100-nm-thick gold film supported on glass, and, typically, the hole diameter ranges from 100–200 nm and the periodicities are between 350 and 700 nm.

Figure 6 shows the white light transmission spectra through two arrays of nanoholes with different periodicities. The experimental setup used in these measurements has been described elsewhere [25,26].

The peaks in the white light transmission correspond to the wavelengths that match the conditions of SPP resonance in each structure. The amount of transmitted light at the peaks far exceeded what was expected from the simple aperture theory [12].

For periodic metallic nanostructures, the phase-matching condition for SP resonance coincides with the Bragg resonances of the grating. At normal incidence, the wavelength of SPP resonance (λ_{SPP}) for a square array of nanoholes can be calculated using [12,27]:

$$\lambda_{\mathrm{SPP}}(i,j) = \frac{p n_{\mathrm{eff}}}{\sqrt{a^2 + b^2}}, \tag{11}$$

where p is the lattice constant (periodicity) of the array, a and b are integers that define the scattering orders of the array and:

$$n_{\mathrm{eff}} = \sqrt{\frac{\varepsilon_{\mathrm{m}}\varepsilon_{\mathrm{d}}}{\varepsilon_{\mathrm{m}} + \varepsilon_{\mathrm{d}}}}. \tag{12}$$

The transmission spectra of arrays at normal incidence presented in Fig. 6 clearly show that the peak resonance depends on the value of p. The asymmetric peak observed in Fig. 6 has been rationalized using a Fano treatment that takes into consideration the interference between two light transmission paths: the incoherent path for direct light transmission through the holes and the coherent path for SPP-assisted transmission [28]. The interference is destructive in the blue side of the resonance peak, producing a small peak followed by a minimum. The interference became constructive after the 180° phase-shift that accompanies the SPP resonance, leading to a sharp increase after the minimum. The effect of the hole-shape on the transmission provides another degree of freedom to the control of light transmission through arrays of nanoholes [29,30]. Experiments with elliptical and square holes demonstrated that the amount of transmitted light at a certain wavelength can be controlled in those cases by the light polarization. In the specific case of rectangular holes, the cut-off condition shifts to larger wavelengths as the width of the rectangular holes are reduced. This is another surprising optical effect observed from nanoholes in metal films [31].

2.2 SPR Sensing Using Arrays of Nanoholes

The simplest way to explore the EOT for chemical applications is on the development of SPR-based sensors. SPR is among the most widely used tools for the investigation of binding events, including affinity tests and other bio-medical applications [2]. The EOT is an SP- based phenomenon, and, according to (11) and (12), the spectral position of the SPR is dependent on the dielectric properties at the interface of the array. The adsorption of molecules on the array shifts the transmission peak wavelength. Figure 7 shows a demonstration of the monolayer sensitivity of the SPR from arrays of nanoholes reported by our group [25,26]. The white light transmission spectrum through a clean array of nanoholes (bare Au surface) presents a distinct resonant peak at 645 nm. The gold surface was then modified by immersing the array of nanoholes in an ethanolic solution of mercaptoundecanoic acid (MUA). A well-organized monolayer of MUA was formed after 24 h

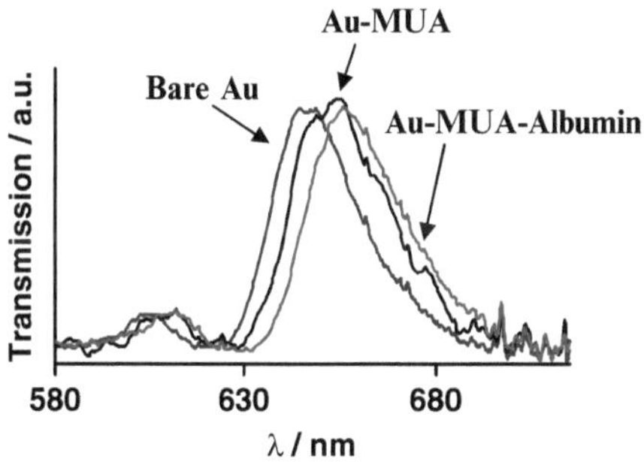

Fig. 7 Effect of molecular adsorption on the EOT. Sequential surface modification with Au-MUA and albumin result in red-shifted spectra. Cleaning restores the surface response to the bare-Au spectrum. Adapted with permission from the American Chemical Society (Brolo et al. 2004)

incubation. The white light spectrum of the array modified by MUA presented a characteristic red-shift in the wavelength of maximum transmission (to 650 nm) due to the changes in the dielectric properties of the surface. Further modification of the surface by a protein (albumin) provoked an additional wavelength shift to 654 nm. The spectrum characteristic of a bare gold surface was recovered after the surface species were removed by a plasma cleaning treatment [25,26].

The sensing demonstration described above was realized in normal transmission, which is much simpler than the typical angled reflection arrangement employed in the commercial SPR systems. This transmission setup is more compatible to the lab-on-chip concept, since miniaturized light sources, such as LEDs, and photodiodes (or a CCD) can sandwich the sensing areas of the array of nanoholes, yielding a compact package.

The output sensitivity observed from arrays of nanoholes in our early report was around 400 nm/RIU (RIU = refractive index units) [25,26]. However, several groups reported increased output sensitivity by exploring new detection schemes and geometries [32–35]. For example, Tetz et al. [35] obtained a 10 times increase in the resolution of the array of nanoholes by implementing a polarization-modulation scheme that changed the Fano line-shape of the SPR to a narrower Lorentzian shape. Lesuffleur et al. [32] also reported increased sensitivity by exploring the additional contribution from localized SP resonance from shaped nanoholes that are used instead of the simple circular apertures. Recent efforts are geared towards multiplexing, achieved by the implementation of 2-D microarrays using the arrays of nanoholes as sensing elements [36–38]. These microarrays allow real-time probing of several binding events in a simplified optical alignment, and they have been used as immunobiosensors [38].

2.3 *Enhanced Spectroscopy Using Arrays of Nanoholes*

A direct consequence of SP excitation is field localization. The SP field is tightly confined to the interface, and it is concentrated in certain areas of nanometric dimensions. This means that the nanostructure enables the focusing of light in subwavelength regions when the conditions for SP excitation are achieved. The spectroscopic response from molecules adsorbed in these regions of high field, known as hot-spots, is expected to be enhanced relative to the signal obtained in the absence of the nanostructure. This effect explains the high sensitivity observed in surface-enhanced Raman scattering (SERS) and surface-enhanced fluorescence spectroscopy (SEFS) [39,40]. The SPP excitation in arrays of nanoholes in noble metals should also lead to field localization and, consequently, enhanced spectroscopy. The near field electric field distribution is predicted to be concentrated around the rims of the holes, providing a well-organized and reproducible distribution of hot-spots under SP resonance conditions [41,42].

Figure 8 shows the surface-enhanced resonance Raman scattering (SERRS) spectra of oxazine 720 adsorbed on arrays of nanoholes in gold films with different

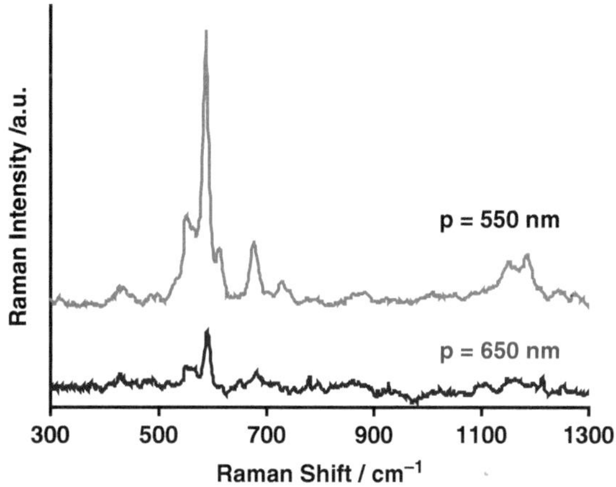

Fig. 8 SERS from oxazine 720 adsorbed on arrays of nanoholes in gold with different periodicities

periodicities. The experiment was realized in forward scattering and the experimental details are given elsewhere [25,26]. The main result from Fig. 8 is that the enhancement of the Raman signal was dependent on the array periodicity. A strong enhancement is observed when the laser energy matches the conditions for SPP excitation.

The best enhancement factor estimated in these experiments was around 10^5 (relative to the normal Raman of liquid benzene). This enhancement factor, however, contains an extra contribution from the resonance Raman (RR) effect. This is because oxazine 720 has an electronic absorption band at 620 nm that is broad enough to match the exciting laser at 632.8 nm. If we consider a typical RR contribution of 10^2, the enhancement factor due to only the nanoholes would be about 10^3. It is important to point out that attempts to obtain the oxazine 720 spectrum from a gold surface without the nanoholes were unsuccessful, indicating that the RR alone cannot explain the increase in the Raman signal. A quantitative evaluation [43] of the enhancement factor from nanoholes in silver showed that the contribution due exclusively to the SPP excitation through nanoholes was around 10^2. SERS from randomly distributed nanoholes in gold films and normal Raman from single nanoholes in metal were also reported [44,45].

Fluorescence-based measurements are already very sensitive and widely used in bio-medical analysis. However, the metallic nanostructures provide further improvement on the sensitivity and limit of detections through the enhancement of the local field. Therefore, a large number of researchers are dedicated to developing substrates for SEFS [46–52]. The effect of the geometrical parameter of the nanostructure on the efficiency of the SEFS is well illustrated in Fig. 9. In this case, the SEFS enhancement factor (SEFS enhancement factor) is plotted against the periodicity of the arrays of nanoholes in gold films. The experiments were realized by spin-coating the arrays of nanoholes with a polystyrene film doped with the oxazine 720 [48].

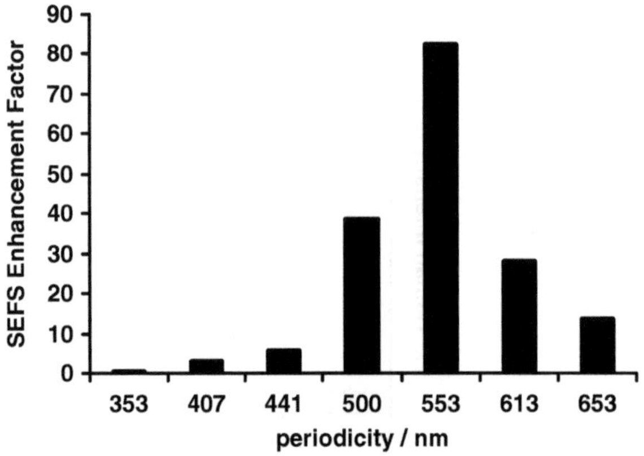

Fig. 9 SEFS enhancement factor calculated for a polystyrene film doped with oxazine 720 on arrays of nanoholes (150 nm diameter) of different periodicities

The amount of fluorescence emission from the arrays was always larger than expected from classical aperture theory when the conditions for SPP resonance were achieved. A maximum enhancement factor of around 80 was observed when the periodicity of the array allows for a SPP excitation that matched the laser energy [48]. This SEFS enhancement factor value is in good agreement with the observations from other groups [51,53]. It is important to mention that the molecules in close contact to the metallic structure, which are the ones that experience the highest field, do not contribute to the SEF, due to the fluorescence quenching experienced by emitters in close proximity to metallic surfaces [54,55]. The application of SEFS from arrays of nanoholes on biosensing has also been demonstrated [50]. The dynamics of biological process from single molecules can be readily investigated, even from relatively high solution concentrations, using nanometric holes [56,57].

Experiments with shaped nanoholes (in contrast to circular holes) offer new possibilities of control over the spectroscopic response of adsorbed species. This is because anisotropic structures present polarization-dependent conditions for SP excitation [29,58–60]. Moreover, apexes in hole structure can further localize the light field, as illustrated in Fig. 10. The "double-hole" structure in Fig. 10 is formed by two nanoholes holes (150 nm diameter) separated by a controllable distance [29,61–65]. The SP resonance depends, then, not only on the same parameters as the circular holes (such as hole diameter and array periodicity), but also on the presence of the apex formed when the two holes touch each other, as shown in Fig. 10c. The apexes act as an optical antenna, focussing the electromagnetic field to a very small region, as shown in Fig. 10d, which is a FDTD calculation of the field distribution around a double-hole with overlapping holes.

It can be seen in Fig. 10d that the electric field is strongly concentrated in the apexes region. Since a strong localized field is the main requirement for enhanced spectroscopy, tunable surface-enhancement is expected from these structures.

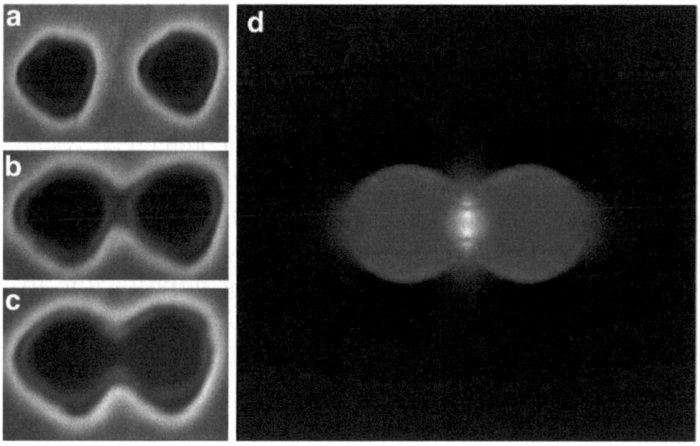

Fig. 10 The double hole structure with different separations between the holes (**a**–**c**), and FDTD calculation (**d**) of the field distribution for the overlapping hole case

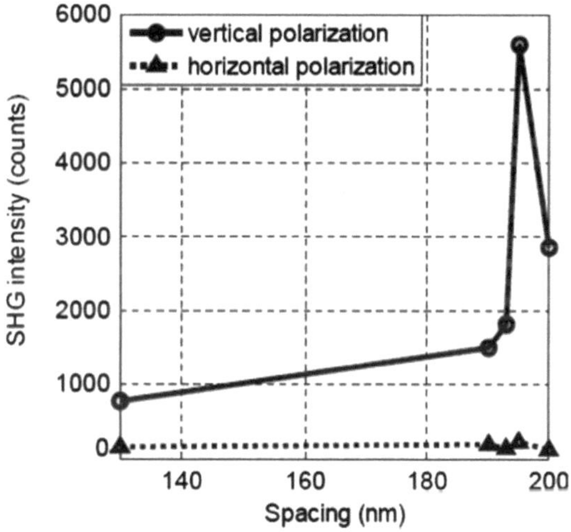

Fig. 11 The dependence of the SHG efficiency on the spacing between the horizontal and vertical polarizations of the fundamental beam. The signal reaches a maximum with overlapped holes (hole diameter is 200 nm). Reprinted with permission from the American Institute of Physics [63]

We have obtained SERS and second harmonic generation (SHG) from the double-hole structures with different hole distances. The dependence of the SHG efficiency on the spacing between the holes in the basis is shown in Fig. 11 for two polarizations of the fundamental beam. It can be seen that, as expected from the calculations shown in Fig. 10d, the SHG signal maximizes when the two holes overlap [63,64]. This optimum distance has also been observed in SERS experiments using the double-hole basis [65]. The double-hole structure offers two degrees of freedom, shape and periodicity, for electric field optimization and enhanced spectroscopy.

3 Integration of Nanohole Sensors

Chemical and biological sensing are of increasing importance in a wide range of applications. Particularly with regard to biosensing, commercial systems providing real-time SPR-based bio-chemical detection capabilities are becoming increasingly common in research laboratories. As a platform, however, this technology is ill-suited to miniaturization. Point-of-care or field applications generally require robust and simple sensing methodologies with supporting infrastructure that facilitates minaturization in concert with micro- and nanofluidics at low cost [66].

Chemical and biological sensing with nanohole arrays are particularly well suited to miniaturization and lab-on-chip integration, owing to several advantages as outlined previously [67]: (1) in contrast to the reflective mode as used in most commercial SPR systems, transmission mode operation at normal incidence simplifies alignment, facilitates the use of high numerical aperture optics, and permits eventual device-level miniaturization and integration of supporting optics; (2) the footprint of a nanohole array is small relative to that required in reflective mode SPR, enabling miniaturization and integration into micro- and nano-fluidic architectures, and higher spatial resolution; (3) in contrast to other local surface plasmon strategies based on, for instance, colloidal nanoparticles or roughened surfaces, nanohole arrays can be fabricated with high reproducibility; (4) the high sensitivity of the optical response to hole geometry, spacing, and orientation provides a large variety of handles with which to tailor the response of the sensor; and (5) the unique geometry differentiating each array element is fixed within the structure and is generally more robust than an adsorbed probe.

Lab-on-chip integration of nanohole array sensors was demonstrated by De Leebeeck et al. [67] through chemical sensing and the monitoring of biomolecule binding using an array of nanohole array sensors in a microfluidic chip platform. Figure 12a, b show schematics of the chip and the optical/fluidic setup employed, respectively. A set of six arrays was employed to spatially resolve a cross-stream concentration gradient, as shown in Fig. 12c. Figure 12d shows the results of the peak shifts resulting in surface-binding in the assembly process of a cysteamine monolayer, biotin linker, and streptavidin protein system.

Pang et al. [33] employed microfluidics to demonstrate the spectral sensititvity of a nanohole-array-based sensor, and achieved sensitivities up to 1520 nm/RIU in calibration tests performed with sequential injection of increasing concentrations of ethylene glycol in water. They also performed real-time monitoring of protein–protein specific binding. With a peak shift detection limit of 0.1 nm, they predict 26 nM anti-BSA could be detected in their system. Lesuffleur et al. [32] employed a flow cell to demonstrate real-time measurements of molecular binding using double-hole nanohole arrays. Sharpe et al. [38] developed a nanohole-array-based immunobiosensor and employed a secondary antibody–gold nanoparticles conjugate to achieve a threefold enhancement in peak shift. This nanoparticle-based enhancement was attributed to cooperative plasmonic effects between the particles and the substrate as both possessed a high dielectric constant relative to the solution.

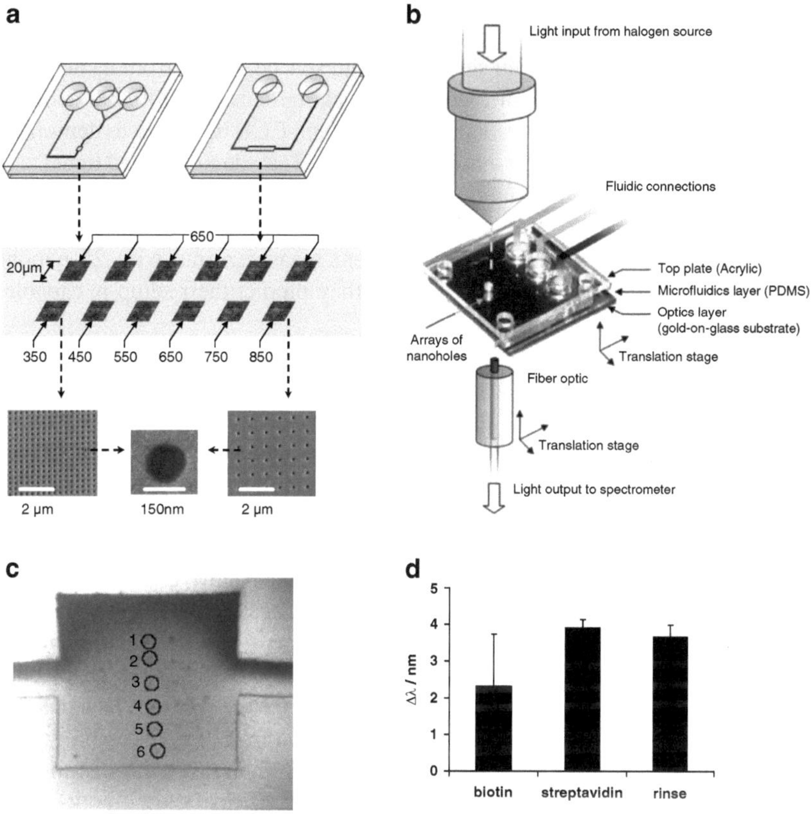

Fig. 12 Lab-on-chip integration and application of arrays of nanohole arrays: (**a**) Schematic of device at range of relevant lengthscales; (**b**) Schematic of optical and fluidic setup with an image of the device; (**c**) Image showing six cross-stream nanohole sensors across a cross-stream microfluidic concentration gradient; and (**d**) Results of on chip biosensing test. Reprinted with permission from the American Chemical Society [67]

Recently, Eftekhari et al. [68] employed rectangular arrays of nanoholes with differing periodicities in orthogonal directions. These biaxial arrays exhibited distinct spectral characteristics depending on the alignment of the polarization of the incident light with the array, and represent an addition handle with which to increase sensitivity, increase probe density, or dynamically tune the spectrum to the incident light in the case of a single-line source.

The use of single-line excitation simplifies the optical measurements as compared to measuring peak shifts in white light transmission. As mentioned earlier in this chapter, laser excitation enables the use of a basic detector, as opposed to a spectrometer, and the change in signal may be attributed directly to the surface condition provided the spectrum exhibits a high slope in the wavelength range surrounding that of the laser source. Lesuffleur et al. [37] employed laser-illumination and CCD detection to monitor the binding kinetics of a self-assembled monolayer of alkanethiolates on gold with a 15-array set of nanohole arrays in parallel. Each 16 × 16 circular nanohole array (200 nm nanohole diameter and 380 nm periodicity) was 6 μm in side length and spaced 50 μm from its neighbour. This arrangement yields a high theoretical packing density of 40,000 spots in a 1 cm^2 area. Ji et al. [36] employed a laser source and CCD detection to demonstrate high-throughput nanohole-array-based biosensing. They demonstrated linear signal response at concentrations ranging from 13–290 nM, and temporal resolution (limited only by camera speed and exposure time) on the order of milliseconds. While Ji and coworkers performed concurrent measurements with 25 nanohole arrays spaced 96 μm apart from each other, they predict their setup is capable of 56 μm spacing.

The near surface nature of SPR limits the extent to which a nanohole-array-based sensor can access the bulk solution. While microfluidic confinement provides some improvement in bulk-to-surface species transport, diffusion rates would be increased greatly in a nano-fluidic channel. Employing a channel with depth on the order of the depth of penetration of the SPR is one such opportunity. In such a configuration, the nanohole array could effectively sample the entire contained volume of analyte. Challenges in such a design include effective fabrication, fluid through-put, and ensuring the structural integrity of such a low aspect ratio channel.

Lab-on-chip integration of nanohole arrays to date has involved a flow-over-array format, that is, the direction of fluid transport is parallel to the substrate surface and orthogonal to the axis of the nanoholes. Diffusion is rapid on the characteristic length of the nanohole, but the ability for the array surface to sample the bulk solution is significantly limited by diffusion in a flow-over-array format. Alternatively, a flow-through-array format, where the fluid flows through the nanoholes between two microfluidic manifolds, would be particularly attractive from a transport perspective; the exposure of the nanoholes to solution would be increased dramatically. Such a flow-through-array format would require additional fabrication considerations, such as ensuring the structural integrity of the film. The strength of the film is particularly important considering the relatively high pressure gradients required to drive fluid through nanoscale holes. Despite some challenges, the benefits of concurrently employing both the photonic and sieving action of the nanoholes are significant. In addition, a flow-through-array geometry could also exploit the double-sided nature of enhanced transmission, that is, the SP generation on both sides of the metallic film.

A schematic of the flow-through nanohole array concept is shown in Fig. 13a. Figure 13b shows computationally predicted biomarker transport within the nanoholes for in-hole average fluid velocities of 1 μm/s and 1 cm/s (as indicated). Reaction rate constants characteristic of surface-based antibody–antigen reactions (with reaction rate constant $k = 10^6$/M/s) [69] were applied at the nanohole walls. For the low average velocity, diffusion of the biomarker (with diffusivity $D = 4\text{x}10^{-11}\ m^2s^{-1}$) to the nanohole surface is effectively complete in one diameter. This result reflects the rapid diffusion characteristic of nanoconfinement. For the higher flow rate case, the absorption of the analyte stream is delayed; however, over 90% bulk adsorption of analyte is attained with the flow rates and nanohole

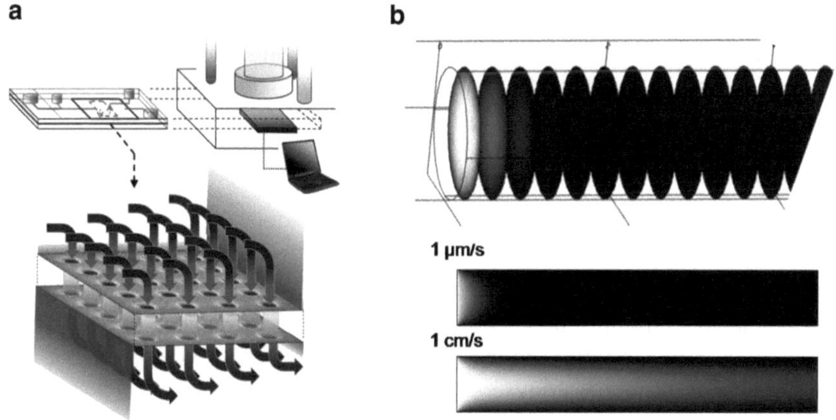

Fig. 13 Integrated flow-through nanohole array concept: (**a**) Schematic of the flow-through nanohole array in a chip-and-reader configuration; (**b**) Results of computational modeling showing predicted biomarker concentration profiles with through-hole fluid average velocities indicated

geometry employed here. This flow rate corresponds to total flow rates (per array) on the order of 2 nL/min, which are compatible with service microfluidics. Implementing the integrated flow-through array format is a focus of our current work.

With respect to integration of nanohole arrays, there are some additional functionalities and combinations of tools that are promising. Firstly, the integration of nanoholes with functionalized beads in a fluid medium presents several opportunities. Microscale and nano-scale beads/particles exhibit a high surface-to-volume ratio and may be readily controlled through flow, electrophoresis, and optical forces. The surface plasmon can, under specific circumstances, be used to selectively trap particles as described elsewhere [70]. With effective particle-control mechanisms, particles with specific surface preparations could serve to concentrate analytes within nanoholes. In addition, potential for cooperative plasmonic effects was demonstrated recently by Sharpe et al. [38], who employed an antibody–gold nanoparticle conjugate to achieve a threefold enhancement in peak shift as compared to that without nanoparticles. Secondly, phase boundaries present an array of coupled optical and fluidic phenomena of potential utility with nanohole arrays. For instance, bubbles in the presence of a wetting liquid provide a mechanism to confine and control liquids to the near-wall region via a film. In addition, capillary effects within nanostructures and optical lensing at gas-liquid interfaces are both topics of fundamental interest [71]. With respect to fluid handling, multiphase microfluidic reactors [72] present several opportunities for the preparation and distribution of solutions over nanohole arrays. Digital microfluidics, or droplet-based microfluidics, in particular appears promising for multiplexed analysis over nano-structured arrays. Thirdly, the integrated flow-through array architecture involves microchannels in series with a set of nanochannels in parallel. Similar configurations have been applied to filtration, as well as to preconcentrate analytes and other functions [73]. Fourthly, the near-surface functionalities of nanohole arrays are synergistic with surface-driven electrokinetic phenomena [74], and the metallic layer could serve as an electrode for a variety of transport mechanisms, including electroosmosis, electrophoresis, and dielectrophoresis.

4 Concluding Remarks

The remarkable optical properties of arrays of subwavelength holes in metal films present many opportunities for chemical and biological sensing, and are well-suited for lab-on-chip integration. The surface-plasmon-mediated electromagnetic field concentration in nanohole arrays enables surface plasmon-resonance-based sensing, as well as surface-enhanced Raman scattering, surface-enhanced fluorescence spectroscopy, and other methodologies. A wide range of chemical and biological sensing applications are possible, given the excellent control over the nano-structure and the optical interactions, provided by now-established nanofabrication and optics. In this chapter, the fundamental phenomena governing optical interactions with metals and metal nanostructures, the sensing methodologies these phenomena enable, and the on-chip integration of nanohole array sensors were discussed. Several integration efforts to date were described here, as also the future prospects for nanohole arrays in a lab-on-chip format and potential to exploit transport phenomena in these structures. Integrated optofluidic nanohole-array-based platforms are an exciting and promising avenue for chemical and biological sensing.

References

1. Maier SA (2007) Plasmonics: fundamentals and applications, 1st edn. Springer, New York
2. Homola J (2008) Surface plasmon resonance sensor for detection of chemical and biological species. Chem Rev 108:462–493
3. Jackson JD (1998) Classical electrodynamics, 3rd edn. Wiley, New York, p 159
4. Haes AJ, Van Duyne RP (2002) A nanoscale optical biosensor: sensitivity and selectivity of an approach based on the localized surface plasmon resonance spectroscopy of triangular silver nanoparticles. J Am Chem Soc 124(35):10596–10604
5. Kneipp K, Wang Y, Kneipp H, Perelman LT, Itzkan I, Dasari RR, Feld MS (1997) Single molecule detection using surface-enhanced Raman scattering (SERS). Phys Rev Lett 78 (9):1667–1670
6. Nie S, Emory SR (1997) Probing single molecules and single nanoparticles by surface-enhanced Raman scattering. Science 275(5303):1102–1106
7. Schaadt DM, Feng B, Yu ET (2005) Enhanced semiconductor optical absorption via surface plasmon excitation in metal nanoparticles. Appl Phys Lett 86(6):063106
8. Bouhelier A, Beversluis M, Hartschuh A, Novotny L (2003) Near-field second-harmonic generation induced by local field enhancement. Phys Rev Lett 90:013903
9. Sipe JE, Boyd RW (2002) Nanocomposite materials for nonlinear optics based on local field effects, in optical properties of nanostructured random media, 82nd edn. Springer, Berlin, pp 1–19

10. Chen CK, de Castro ARB, Shen YR (1981) Surface-enhanced second-harmonic generation. Phys Rev Lett 46:145–148
11. van Nieuwstadt JAH, Sandtke M, Harmsen RH, Segerink FB, Prangsma JC, Enoch S, Kuipers L (2006) Strong modification of the nonlinear optical response of metallic subwavelength hole arrays. Phys Rev Lett 97:146102
12. Ebbesen TW, Lezec HJ, Ghaemi HF, Thio T, Wolff PA (1998) Extraordinary optical transmission through sub wavelength hole arrays. Nature 391:667–669
13. Bethe HA (1944) Theory of diffraction by small holes. Phys Rev 66:163–182
14. Wannemacher R (2001) Plasmon-supported transmission of light through nanometric holes in metallic thin films. Opt Commun 195:107–118
15. Degiron A, Lezec HJ, Yamamoto N, Ebbesen TW (2004) Optical transmission properties of a single subwavelength aperture in a real metal. Opt Commun 239:61–66
16. García de Abajo FJ, Gómez-Medina G, Sáenz JJ (2005) Full transmission through perfect-conductor subwavelength hole arrays. Phys Rev E 72:016608
17. Gordon R (2007) Bethe's aperture theory for arrays. Phys Rev A 76:053806
18. Gay G, Alloschery O, De Lesegno BV, O'Dwyer C, Weiner J, Lezec HJ (2006) The optical response of nanostructured surfaces and the composite diffracted evanescent wave model. Nat Phys 2:262–267
19. Lezec HJ, Thio T (2004) Diffracted evanescent wave model for enhanced and suppressed optical transmission through subwavelength hole arrays. Opt Express 12:3629–3651
20. Flammer PD, Schick IC, Collins RT, Hollingsworth RE (2007) Interference and resonant cavity effects explain enhanced transmission through subwavelength apertures in thin metal films. Opt Express 15:7984–7993
21. Gao HW, Henzie J, Odom TW (2006) Direct evidence for surface plasmon-mediated enhanced light transmission through metallic nanohole arrays. Nano Lett 6:2104–2108
22. Garcia-Vidal FJ, Martin-Moreno L (2002) Transmission and focusing of light in one-dimensional periodically nanostructured metals. Phys Rev B 66:155412
23. Huang CP, Wang QJ, Zhu YY (2007) Dual effect of surface plasmons in light transmission through perforated metal films. Phys Rev B 75:245421
24. Popov E, Neviere M, Enoch S, Reinisch R (2000) Theory of light transmission through subwavelength periodic hole arrays. Phys Rev B 62:16100–16108
25. Brolo AG, Arctander E, Gordon R, Leathem B, Kavanagh KL (2004) Nanohole-enhanced Raman scattering. Nano Lett 4:2015–2018
26. Brolo AG, Gordon R, Leathem B, Kavanagh KL (2004) Surface plasmon sensor based on the enhanced light transmission through arrays of nanoholes in gold films. Langmuir 20: 4813–4815
27. Ghaemi HF, Thio T, Grupp DE, Ebbesen TW, Lezec HJ (1998) Surface plasmons enhance optical transmission through subwavelength holes. Phys Rev B 58:6779–6782
28. Fano U (1961) Effects of configuration interaction on intensities and phase shifts. Phys Rev 124:1866–1878
29. Gordon R, Hughes M, Leathem B, Kavanagh KL, Brolo AG (2005) Basis and lattice polarization mechanisms for light transmission through nanohole arrays in a metal film. Nano Lett 5:1243–1246
30. van der Molen KL, Klein KKJ, Enoch S, Segerink FB, van Hulst NF, Kuipers L (2005) Role of shape and localized resonances in extraordinary transmission through periodic arrays of subwavelength holes: experiment and theory. Phys Rev B 72(045421):1–9
31. Gordon R, Brolo AG (2005) Increased cut-off wavelength for a subwavelength hole in a real metal. Opt Express 13:1933–1938
32. Lesuffleur A, Im H, Lindquist NC, Oh SH (2007) Periodic nanohole arrays with shape-enhanced plasmon resonance as real-time biosensors. Appl Phys Lett 90:261104
33. Pang L, Hwang GM, Slutsky B, Fainman Y (2007) Spectral sensitivity of two-dimensional nanohole array surface plasmon polariton resonance sensor. Appl Phys Lett 91:123112

34. Stark PRH, Halleck AE, Larson DN (2005) Short order nanohole arrays in metals for highly sensitive probing of local indices of refraction as the basis for a highly multiplexed biosensor technology. Methods 37:37–47
35. Tetz KA, Pang L, Fainman Y (2006) High-resolution surface plasmon resonance sensor based on linewidth-optimized nanohole array transmittance. Opt Lett 31:1528–1530
36. Ji J, O'Connell JG, Carter DJD, Larson DN (2008) High-throughput nanohole array based system to monitor multiple binding events in real time. Anal Chem 80:2491–2498
37. Lesuffleur A, Im H, Lindquist NC, Lim KS, Oh SH (2008) Laser-illuminated nanohole arrays for multiplex plasmonic microarray sensing. Opt Express 16:219–224
38. Sharpe JC, Mitchell JS, Lin L, Sedoglavich H, Blaikie RJ (2008) Gold nanohole array substrates as immunobiosensors. Anal Chem 80:2244–2249
39. Lakowicz JR (2006) Plasmonics in biology and plasmon-controlled fluorescence. Plasmonics 1:5–33
40. Moskovits M (1985) Surface-enhanced spectroscopy. Rev Mod Phys 57:783–826
41. Chang SH, Gray SK, Schatz GC (2005) Surface plasmon generation and light transmission by isolated nanoholes and arrays of nanoholes in thin metal films. Opt Express 13:3150–3165
42. Krishnan A, Thio T, Kima TJ, Lezec HJ, Ebbesen TW, Wolff PA, Pendry J, Martin-Moreno L, Garcia-Vidal FJ (2001) Evanescently coupled resonance in surface plasmon enhanced transmission. Opt Commun 200:1–7
43. Reilly TH, Chang SH, Corbman JD, Schatz GC, Rowlen KL (2007) Quantitative evaluation of plasmon enhanced Raman scattering from nanoaperture arrays. J Phys Chem C 111: 1689–1694
44. Bahns JT, Yan FN, Qiu DL, Wang R, Chen LH (2006) Hole-enhanced Raman scattering. Appl Spectrosc 60:989–993
45. Wenger J, Dintinger J, Bonod N, Popov E, Lenne PF, Ebbesen TW, Rigneault H (2006) Raman scattering and fluorescence emission in a single nanoaperture: optimizing the local intensity enhancement. Opt Commun 267:224–228
46. Blair S, Chen Y (2001) Resonant-enhanced evanescent-wave fluorescence biosensing with cylindrical optical cavities. Appl Opt 40:570–582
47. Brolo AG, Kwok SC, Cooper MD, Moffitt MG, Wang CW, Gordon R, Riordon J, Kavanagh KL (2006) Surface plasmon-quantum dot coupling from arrays of nanoholes. J Phys Chem B 110:8307–8313
48. Brolo AG, Kwok SC, Moffitt MG, Gordon R, Riordon J, Kavanagh KL (2005) Enhanced fluorescence from arrays of nanoholes in a gold film. J Am Chem Soc 127:14936–14941
49. Kim JH, Moyer PJ (2007) Laser-induced fluorescence within subwavelength metallic arrays of nanoholes indicating minimal dependence on hole periodicity. Appl Phys Lett 90:131111
50. Liu Y, Bishop J, Willians L, Blair S, Herron J (2004) Biosensing based upon molecular confinement in metallic nanocavity arrays. Nanotechnology 15:1368–1374
51. Liu YD, Blair S (2003) Fluorescence enhancement from an array of subwavelength metal apertures. Opt Lett 28:507–509
52. Stark PRH, Halleck AE, Larson DN (2007) Breaking the diffraction barrier outside of the optical near-field with bright, collimated light from nanometric apertures. Proc Natl Acad Sci USA 104:18902–18906
53. Garrett SH, Smith LH, Barnes WL (2005) Fluorescence in the presence of metallic hole arrays. J Mod Opt 52:1105–1122
54. Lakowicz JR, Shen YB, D'Auria S, Malicka J, Fang JY, Gryczynski Z, Gryczynski I (2002) Radiative decay engineering. 2. Effects of silver island films on fluorescence intensity, lifetimes, and resonance energy transfer. Anal Biochem 301:261–277
55. Ritchie G, Burstein E (1981) Luminescence of dye molecules adsorbed at a Ag surface. Phys Rev B 24:4843–4846
56. Levene MJ, Korlach J, Turner SW, Foquet M, Craighead HG, Webb WW (2003) Zero-mode waveguides for single-molecule analysis at high concentrations. Science 299:682–686

57. Wenger J, Lenne PF, Popov E, Rigneault H, Dintinger J, Ebbesen TW (2005) Single molecule fluorescence in rectangular nano-apertures. Opt Express 13:7035–7044
58. Degiron A, Ebbesen TW (2005) The role of localized surface plasmon modes in the enhanced transmission of periodic subwavelength apertures. J Opt A Pure Appl Opt 7:S90–S96
59. Gordon R, Brolo AG, McKinnon A, Rajora A, Leathem B, Kavanagh KL (2004) Strong polarization in the optical transmission through elliptical nanohole arrays. Phys Rev Lett 92:037401
60. Koerkamp KJK, Enoch S, Segerink FB, Hulst NFV, Kuipers L (2004) Strong influence of hole shape on extraordinary transmission through periodic arrays of nanoholes. Phys Rev Lett 92:183901
61. Kumar LKS, Gordon R (2006) Overlapping double-hole nanostructure in a metal film for localized field enhancement. IEEE J Sel Top Quantum Electron 12:1228–1232
62. Kumar LKS, Lesuffleur A, Hughes MC, Gordon R (2006) Double nanohole apex-enhanced transmission in metal films. Appl Phys B 84:25–28
63. Lesuffleur A, Kumar LKS, Gordon R (2006) Enhanced second harmonic generation from nanoscale double-hole arrays in a gold film. Appl Phys Lett 88(26):261104
64. Lesuffleur A, Kumar LKS, Gordon R (2007) Apex-enhanced second-harmonic generation by using double-hole arrays in a gold film. Phys Rev B 75:045423
65. Lesuffleur A, Kumar LKS, Brolo AG, Kavanagh KL, Gordon R (2007b) Apex-enhanced raman spectroscopy using double-hole arrays in a gold film. J Phys Chem C 111:2347–2350
66. Chin CD, Linder V, Sia SK (2007) Lab-on-a-chip devices for global health. Lab Chip 7:41–57
67. De Leebeeck A, Kumar LKS, de Lange V, Sinton D, Gordon R, Brolo AG (2007) On-chip surface-based detection with nanohole arrays. Anal Chem 79:4094–4100
68. Eftekhari F, Gordon R, Ferreira J, Brolo AG, Sinton D (2008) Polarization-dependent sensing of a self-assembled monolayer using biaxial nanohole arrays. Appl Phys Lett 92:253103
69. Wofsy C, Goldstein B (2002) Effective rate models for receptors distributed in a layer above a surface: application to cells and biacore. Biophys J 82:1743–1755
70. Sinton D, Gordon R, Brolo AG (2008) Nanohole arrays in metal films as optofluidic elements: progress and potential. Microfluid Nanofluidics 4:107–116
71. Eijkel C, van den Berg TA (2005) Nanofluidics: what is it and what can we expect from it? Microfluid Nanofluidics 1:249–267
72. Günther A, Jensen KJ (2006) Multiphase microfluidics: from flow characteristics to chemical and materials synthesis. Lab Chip 6:1487–1503
73. Abgrall P, Nguyen NT (2008) Nanofluidic devices and their applications. Anal Chem 80:2326–2341
74. Squires TM, Quake SR (2005) Microfluidics: fluid physics at the nanoliter scale. Rev Mod Phys 77:977–1026

Nanostructure-Based Localized Surface Plasmon Resonance Biosensors

Donghyun Kim

Abstract This chapter reviews the characteristics of localized surface plasmon resonance (LSPR), the excitation of which is mediated by nanostructures, and its applications to biosensing. The LSPR is explored in three regimes in terms of creation and coupling of localized surface plasmons (LSPs): LSPs created in surface-relief patterns coupled to propagating surface plasmons (SPs), LSPs in surface-relief patterns coupled to particle plasmons, and LSPs created in particles. The results, in general, suggest that localized field enhancement in the near-field be correlated with enhanced detection sensitivity for LSPR over conventional thin film-based SP resonance while LSPR-based biosensors can potentially maintain flexibility by using nanoparticles.

Keywords Nanostructures · Surface plasmon resonance · Localized surface plasmon resonance · Bio-molecular interactions · Refractive index change · Effective medium · Thin films · Biosensors · Sensitivity · Nanoparticles

Contents

D. Kim
School of Electrical and Electronic Engineering, Yonsei University, Seoul 120-749, Korea
e-mail: kimd@yonsei.ac.kr

M. Zourob and A. Lakhtakia (eds.), *Optical Guided-wave Chemical and Biosensors 1*, Springer Series on Chemical Sensors and Biosensors 7,
DOI 10.1007/978-3-540-88242-8_7,

Abbreviations

CCD	Charge coupled device
DNA	Deoxyribonucleic acid
EMT	Effective medium theory
FOM	Figure of merit
HDT	1,6-hexanedithiol
LSP	Localized surface plasmon
LSPR	Localized surface plasmon resonance
SAM	Self-assembled monolayer
SEF	Sensitivity enhancement factor
SP	Surface plasmon
SPR	Surface plasmon resonance
VF	Volume fraction

Symbols

c	Free-space light speed
d_{f}	Film thickness
d_{NW}	Nanowire thickness
d_{SAM}	Thickness of a self-assembled monolayer
e	Electron charge
f	Fill factor
k_0	Free-space light wave vector
k_{SP}	Plasmon momentum
m	Electron mass
n_{e}	Electron number density
n_{env}	Ambient refractive index
n_{SAM}	Refractive index of a self-assembled monolayer
R_{min}	Minimum reflectance at resonance
S	Slope of the resonance angle or resonance wavelength
T_{max}	Maximum transmittance at resonance
ω	Light angular frequency
w_{NW}	Nanowire width
w_{p}	Plasma angular frequency
ε_{d}	Dielectric permittivity
$\varepsilon_{\mathrm{eff}}$	Effective permittivity
ε_{m}	Metal permittivity
$\varepsilon_{\mathrm{np}}$	Nanoparticle permittivity
ε_{A} (ε_{B})	Permittivity of material A (B)
$\varepsilon_{0,\mathrm{TE}}$	Zeroth-order effective permittivity for TE polarization

$\varepsilon_{0,TM}$	Zeroth-order effective permittivity for TM polarization
θ_{SP}	Plasmon resonance angle
θ_{LSP}	Localized plasmon resonance angle
Λ	Period
λ	Light wavelength

1 Introduction

Surface plasmon resonance (SPR) is one of the commercially successful optical sensing techniques. The conception of SPR dates back to the 1900s when R.M. Wood first noticed, in 1902, anomalous dark and light bands of light reflected by a metal backed diffraction grating and Lord Rayleigh theoretically treated the anomaly in 1907 [1, 2]. As a sensing technique, SPR was initially used for thin film characterization in 1970s. At the turn of 1980s, it was applied for the first time to characterizing bio-molecular interactions [3, 4]. With a few key techniques, this laid the basis for founding Biacore AB (now, a GE Healthcare company), one of the primary manufacturers in the optical sensor system market, which released the first commercial SPR instrument [5].

While it is safe to say that SPR is a mature technique from the historical perspective, new driving forces appear to challenge traditional SPR for various needs that traditional SPR sensors fail to satisfy. In particular, a novel SPR biosensor that attempts to capitalize on the nanotechnology, by which to localize surface plasmons (SPs), has emerged and thus has been appropriately called a localized surface plasmon resonance (LSPR) biosensor. In this chapter, I focus on the LSPR biosensor by reviewing its operating principles and properties in a systematic way and venture into future directions along which LSPR biosensors evolve.

Before I begin LSPR biosensors, let me overview briefly how SPR works. For details of SPR, readers are advised to consult a related review, for example, [6]. Obviously, SPs are highly excited at SPR. SP refers to a longitudinal electron concentration wave formed at a metal–dielectric interface. In other words, initially uniform free electrons in metal form concentration waves at surface, similar to pressure waves in air. Electron nonuniformity is why it is called plasmon and the induced dipoles produce extremely thin surface evanescent waves in the range of 100–200 nm. For LSPR, the induced dipole distribution changes according to specific localization, which also affects spatial profiles of evanescent field amplitudes. SPR is established only when the incident light is p-polarized, i.e., its magnetic field oscillates in the direction that is contained in the surface plane. Note that SPs may be created with s-polarized light in the case that dielectric films are deposited on metallic surface, which can be employed to characterize the anisotropy of target molecules [7]. The dispersion relation between SP wave number k_{SP} and incident light angular frequency ω can be calculated from Maxwell's equations and is given by

$$k_{SP} = \frac{\omega}{c}\sqrt{\frac{\varepsilon_m \varepsilon_d}{\varepsilon_m + \varepsilon_d}} = k_0 \sin\theta_{SP}, \quad (1)$$

where θ, ε_m, and ε_d represent the angle of incidence, metal permittivity, and dielectric permittivity, respectively. The symbols c and k_0 stand for the speed of light and wave number in the free-space, respectively. For derivation and other details of (1), see [8]. What one measures using SPR is basically the change in ε_d in (1) as an interaction between biomolecules occurs.

The sensitivity limit of conventional SPR biosensing is typically quoted as 1 pg/mm^2, although routinely achieved sensitivity would be close to 1 ng/mm^2 [9]. In terms of relatively heavy biomolecules (~100 kDa), 1 pg/mm^2 amounts to detecting six million molecules per mm^2. Compared with other optical sensing techniques, for example, use of fluorescent dyes, SPR suffers from relatively low sensitivity. This stems from the very fact that SPR is a nonlabeling technique, often a significant advantage over labeling techniques such as using fluorescence.

There are four different ways of implementing an SPR setup: measuring intensity changes by scanning angles or wavelength of incidence (the latter thus being sometimes called SPR spectroscopy), measuring reflectivity changes directly which is called imaging SPR or SPR microscopy, and finally measuring phase changes at SPR. In terms of sensitivity, which setup performs the best is often a question of intense debate [10]. A correct answer to which is the best or provides the highest sensitivity really depends on who you talk to and it seems that the sensitivity performance depends on specific applications. However, it may be that an angle scanning setup shows the highest sensitivity, considering the optical components available and the ease of implementation altogether. A wavelength scanning setup may show comparably high sensitivity, although it requires extensive calibration. While a phase measurement setup has also been used with high sensitivity, it suffers from insufficient dynamic range, not to mention the difficulty associated with measuring phase changes during an interaction. Imaging SPR probably shows the lowest sensitivity, in which case the sensitivity is limited by the performance of a detector, oftentimes CCD. Imaging SPR, however, has a very strong advantage in high-throughput sensing, and thus is appropriate for applications that require high-throughput measurement at not so high sensitivity.

2 LSPR Biosensors

Now, I turn to LSPR excited by nanostructures in an order of reduced coupling of LSPs to propagating SPs, that is, based on (1) excitation of LSPs through surface relief nanostructures, (2) coupling between particle plasmon and LSP, and (3) particle plasmons.

2.1 LSPR Biosensor Based on Surface Relief Nanostructures

Surface relief nanostructures may be used to create LSPs. In contrast to using particles to excite plasmons, surface relief patterns have an advantage of no aggregation and thus offer reliable sensing performance. Much part of the discussion addresses field enhancement in the near-field. Near-field enhancement can lead to amplification of signals produced by bio-molecular interactions near surface.

2.1.1 Geometry

As a representative geometry of a surface relief nanostructure of interest, let us focus on a one-dimensional nanograting or periodic nanowire arrays shown in Fig. 1.

The simple structure, in addition to facilitating computation of optical properties, may allow understanding specific interactions between LSPs and propagating SPs. However, it is reasonably complicated to provide many design parameters that one can play with: Λ (period), d_{NW} (nanowire depth), f (fill factor), profile, grating material, d_f (film thickness), and film material. For convenience, I fix d_f at 40 nm. Note that a Cr layer is intended for good attachment of gold films to a dielectric substrate and usually degrades optical characteristics. Self-assembled monolayers (SAMs) in Fig. 1 represent target interactions.

2.1.2 Dispersion Relation

A dispersion relation describes electromagnetic wave radiation created by distributed electron dipole distribution that is associated with a given nanostructure on surface. The distribution of electrons is extremely sensitive to structural changes. The dispersion relation of LSPs is not known in a simple form. However, one may obtain a closed form similar to (1) for LSPs excited in a periodic structure using an effective medium theory (EMT). A second-order EMT by Rytov states that an inhomogeneous layer of two alternating materials (n_A and n_B representing

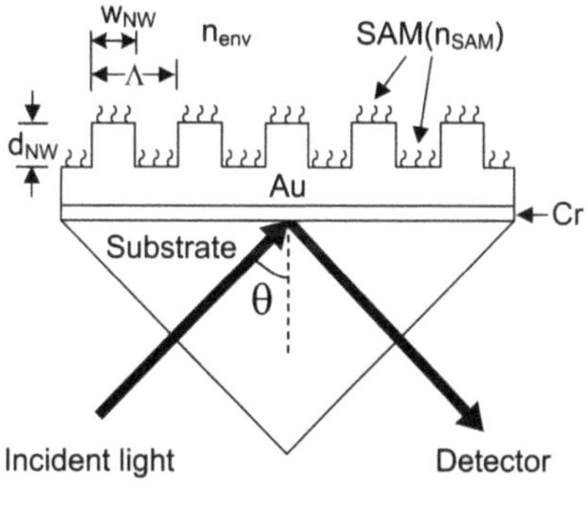

Fig. 1 Schematic of an LSPR biosensor based on periodic nanowires

their refractive indices) with a period Λ may be replaced with a homogeneous anisotropic layer if $\Lambda << \lambda$ [11], such that

$$\varepsilon_{\rm eff} = \varepsilon_{0,\rm TM} + \frac{\pi^2}{3} f^2 (1-f)^2 \left(\frac{1}{\varepsilon_{\rm A}} - \frac{1}{\varepsilon_{\rm B}}\right)^2 \varepsilon_{0,\rm TM}{}^3 \varepsilon_{0,\rm TE} \left(\frac{\Lambda}{\lambda}\right)^2, \tag{2}$$

where f is the grating fill factor and $\varepsilon_{\rm A}$ and $\varepsilon_{\rm B}$ are relative permittivities of the grating materials. The zeroth-order permittivity ε_0 in (2) is given by

$$\begin{aligned} \varepsilon_{0,\rm TE} &= f\varepsilon_{\rm A} + (1-f)\varepsilon_{\rm B}, \\ \varepsilon_{0,\rm TM} &= \frac{\varepsilon_{\rm A}\varepsilon_{\rm B}}{f\varepsilon_{\rm B} + (1-f)\varepsilon_{\rm A}}. \end{aligned} \tag{3}$$

While $\varepsilon_{\rm eff}$ is a complex number, it can be dominantly negative real, depending on f. For the structure shown in Fig. 1, the dispersion relation for nonmagnetic materials can be given by

$$k_{\rm SP} = \frac{\omega}{c}\sqrt{\frac{\varepsilon_{\rm m}\varepsilon_{\rm eff}}{\varepsilon_{\rm m} + \varepsilon_{\rm eff}}}. \tag{4}$$

It was found that (4) agrees fairly well to describe near-field field distribution except for the regions of strong LSPR, as shown in Fig. 2 [12].

In the case of $|\varepsilon_{\rm m}'| >> \varepsilon_{\rm m}''$, the complex term $k_{\rm SP}$ is taken as

$$\mathrm{Re}(k_{\rm SP}) \approx \frac{\omega}{c}\left(\frac{\varepsilon_{\rm m}'\varepsilon_{\rm eff}'}{\varepsilon_{\rm m}' + \varepsilon_{\rm eff}'}\right)^{1/2}, \tag{5}$$

$$\mathrm{Im}(k_{\rm SP}) \approx \frac{\omega}{c}\left(\frac{\varepsilon_{\rm m}'\varepsilon_{\rm eff}'}{\varepsilon_{\rm m}' + \varepsilon_{\rm eff}'}\right)^{3/2}, \frac{\varepsilon_m{}''}{2(\varepsilon_m{}')^2}, \tag{6}$$

where $\varepsilon_{\rm eff}''$ is assumed to be negligible. The enhancement of LSPR can be ascribed to the effective permittivity of the nanowire layer approaching that of gold in magnitude, i.e. $|\varepsilon_{\rm m}'| \sim \varepsilon_{\rm eff}'$. Now, the dispersion relation can be obtained by inserting (2) and (3) into (5). Here, let me assume that $\varepsilon_{\rm eff}'$ is not dispersive, although the procedure can be a lot more complicated if a Drude model is taken for $\varepsilon_{\rm eff}'$. With

$$\varepsilon_{\rm m} = 1 - \left(\frac{\omega_{\rm p}}{\omega}\right)^2 \tag{7}$$

and $\omega_{\rm p} = (4\pi n_{\rm e}e^2/m)^{1/2}$, where $n_{\rm e}$ is the electron number density, e electron charge, and m electron mass, and for $ck_{\rm SP} << \omega_{\rm p}$,

$$\omega \approx ck_{\rm SP}. \tag{8}$$

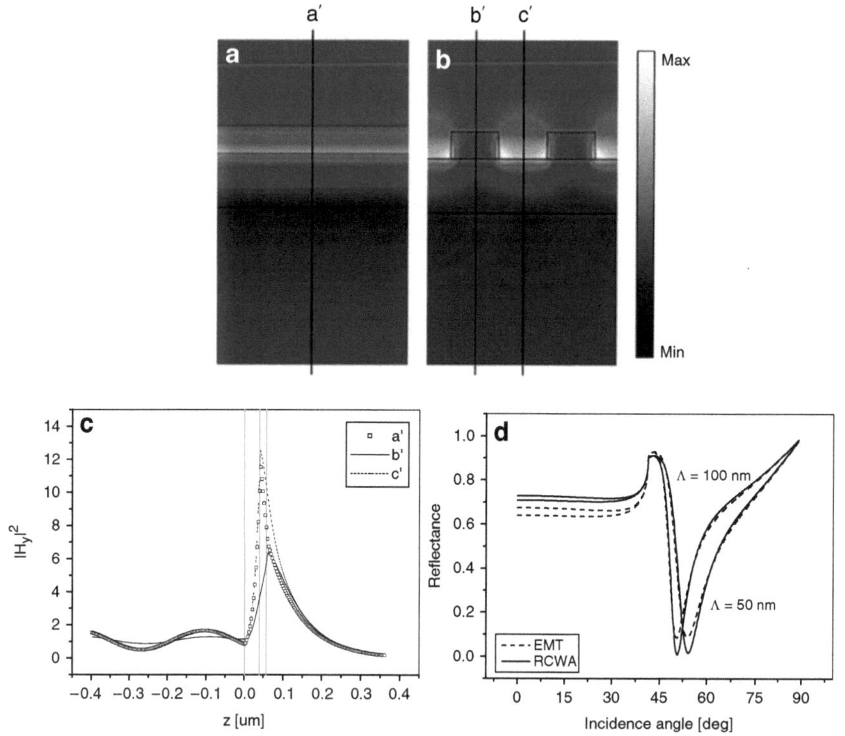

Fig. 2 Magnetic field intensity ($|Hy|^2$) of a nanowire LSPR structure at VF = 10% and $\Lambda = 50$ nm (VF: volume factor equivalent to a fill factor): (**a**) by the EMT based on a four-layer Fresnel's equation given in (1) and (2) and (**b**) near-field distribution calculated by FDTD; (**c**) field profiles by EMT (across a cross-section represented by a′) and FDTD (b′ through a nanowire and c′ between nanowires). (**d**) SPR characteristics of a nanowire LSPR structure with $d_f = 40$ nm and $d_g = 20$ nm, calculated by rigorous coupled wave analysis (solid) and EMT (dotted) when VF = 50%. Reprinted from [12], Copyright (2007), with permission from the Optical Society of America

However, if $ck_{SP} >> \omega_p$,

$$\omega \approx \frac{\omega_p}{\sqrt{1+\varepsilon_{eff}}}. \tag{9}$$

The dispersion curves, shown in Fig. 3 based on (8) and (9), depict the LSP excitation in a periodic nanostructure qualitatively. The variation of plasmon momentum at momentum-matching can be described with ε_{eff} of a nanostructure in (9). For example, the second-order term of (2) is negative, which can decrease ε_{eff} to be smaller than ε_d, with ε_d as the permittivity of a dielectric substrate. In this case, Fig. 3 shows that plasmon is momentum enhanced by using nanostructures.

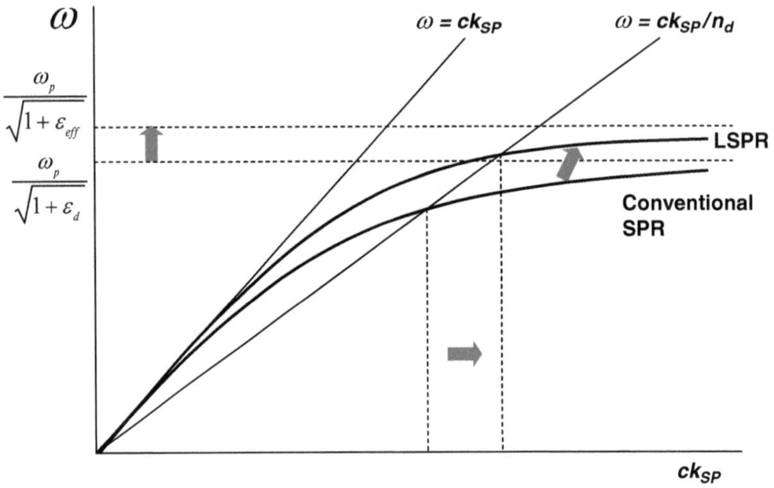

Fig. 3 Dispersion relation for SPs. The *arrow* indicates the change of a dispersion relation of LSPR relative to conventional SPR

2.1.3 Figures of Merit

As a quantitative measure of the far-field performance, two quantities may be introduced. First of all, sensitivity enhancement factor (SEF), a ratio of resonance angle shift due to target analyte binding on a nanowire-based biosensor to that of a conventional SPR structure using a thin gold film with equal thickness, is defined as

$$\mathrm{SEF} = \left|\frac{\delta\theta_{\mathrm{LSP}}}{\delta\theta_{\mathrm{SP}}}\right| = \left|\frac{\theta_{\mathrm{LSP}}(\text{target analyte}) - \theta_{\mathrm{LSP}}(\text{no analyte})}{\theta_{\mathrm{SP}}(\text{target analyte}) - \theta_{\mathrm{SP}}(\text{no analyte})}\right|, \tag{10}$$

where θ_{LSP} and θ_{SP} represent the resonance angles with and without nanowires, respectively. An SEF of (10) is defined as a large-signal quantity that references "no analyte" in (10) as pure water in the case of sensing an environmental change or as no SAM when a biosensor detects a layer formation. An SEF can also be defined as a differential SEF that simulates the differential sensitivity enhancement in response to relatively small perturbations about the refractive index prior to or no target interactions. For a biosensor, the differential SEF would be appropriate on the grounds that bio-molecular changes are often associated with a very small change in refractive indices, e.g., $\delta n \approx 0.03$ from pure water to pure ethanol.

Also note that the specific form of (10) addresses the sensitivity improvement in the angle-scanning scheme. It can be easily extended for other schemes. For example, SEF may be redefined for SPR imaging as

$$\mathrm{SEF} = \left|\frac{\delta R_{\mathrm{LSP}}}{\delta R_{\mathrm{SP}}}\right| = \left|\frac{R_{\mathrm{LSP}}(\text{target analyte}) - R_{\mathrm{LSP}}(\text{no analyte})}{R_{\mathrm{SP}}(\text{target analyte}) - R_{\mathrm{SP}}(\text{no analyte})}\right|, \tag{11}$$

where R refers to reflectance in the corresponding configuration.

Secondly, a figure of merit (FOM) can be defined to effectively compare the overall performance of optical sensors as

$$\mathrm{FOM} = \frac{S}{\mathrm{FWHM}} \cdot \frac{1}{R_{\min}}, \tag{12}$$

where S is the slope of the resonance angle or resonance wavelength over the refractive-index range, which corresponds to the sensitivity of an SPR sensor. In our case, S is equal to the differential SEF. The FOM is equal to the quality factor that takes into account the reflectance for a reflection-type SPR structure. A smaller FWHM and R_{min} are desired because a deeper and narrower resonance peak allows efficient detection of the resonance shift and precise analysis of sensing events. For transmission-based SPR detection, $1/R_{\min}$ can be changed to $T_{\max}$ in (12) [13].

2.1.4 Effects of Structural Parameters

2.1.4.1 Grating Material

For excellent biocompatibility and wide availability of linker molecules, gold is the most popular material for metallic films and surface-relief nanostructures. Oftentimes, silver provides improved sensing performance, although it is somewhat of less practical importance, because of the reactive and tarnishing nature of silver. Aluminum may also be a possibility to consider. However, its oxidizing tendency makes it difficult to develop stable linker layers for various targets.

2.1.4.2 Period and Fill Factor

One may be gullible to say that a small nanowire period is always desired to induce local plasmonic field enhancement, thereby strong sensitivity enhancement. What may really be critical is net nanowire or nano-groove width, i.e., $f\Lambda$ or $(1-f)\Lambda$. In other words, even with a large nanowire period, one may excite LSPs efficiently if a fill factor is sufficiently small. There seems to exist a fill factor value at which local fields are maximally induced. Figure 4 presents resonance angles with and without a 1-nm thick 1,6-hexanedithiol (HDT) dielectric SAM as a VF is varied and shows enhanced plasmon momentum at $\Lambda = 50$ nm and VF ~0.8 and also at $\Lambda = 100$ nm and VF ~ 0.9, when nanowire gap is 10 nm in both cases [14]. In other words, absolute nanowire gap may be more important in plasmon enhancement than nanowire period. These trends have been reported in many near-field studies [15, 16].

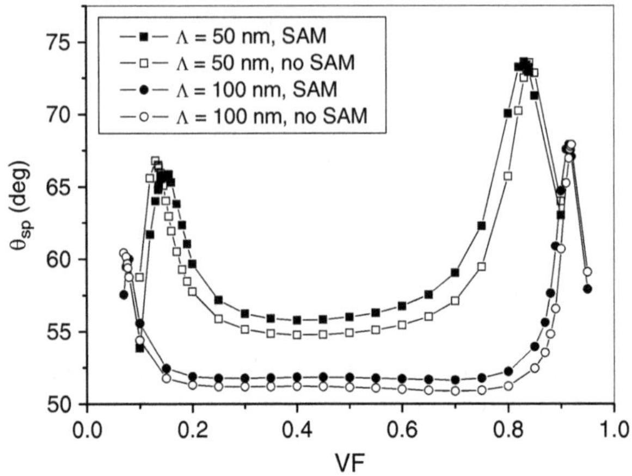

Fig. 4 Variation of θ_{SP} with VF at $\Lambda = 50$ and 100 nm in air, assuming with and without a 1-nm thick 1,6-hexanedithiol dielectric SAM. Reprinted from [14], Copyright (2006), with permission from the Optical Society of America

Note that Fig. 4 was produced for an air environment. In water ambience, the enhancement of plasmon momentum is expected to be less drastic because of reduced index contrast.

2.1.4.3 Depth

Depth can be an indicator of interactions between LSPs and bulk SPs. In thick nanowires, an LSP is well isolated from bulk SPs, while LSPs interact with and are affected from bulk SPs significantly in shallow nanowires. It was shown that penetration depth was calculated to be smaller for thicker nanowires, as a result of increased field localization [17]. In fact, a nanowire period (Λ) and a fill factor (f) should also be considered in this picture, in the sense that LSPs tend to be coupled to each other at appropriate values of Λ and f. On the one hand, if LSPs perturb SPs weakly, i.e., LSPs are dominated by SPs, local field enhancement and sensitivity improvement over conventional SPR structure are rather weak. On the other hand, if LSPs dominate SPs, resonance characteristics become so broad that resonance effectively disappears. In short, an optimum nanowire depth exists, although the optimum may depend on other parameters.

Figure 5 represents the plasmon momentum with and without a 1-nm thick HDT-SAM in air ambience as VF is varied for various combinations of nanowire periods and depths [18]. VF is the volume ratio of metal and is equal to f for one-dimensional grating. First of all, trends of enhanced LSPR with smaller Λ and larger d_{NW} are obvious, because increased isolation of a LSP mode encourages decoupling from a bulk SP mode, as was observed in earlier studies on SPs created on gold and silver gratings [19, 20]. Potential for large sensitivity increase of a

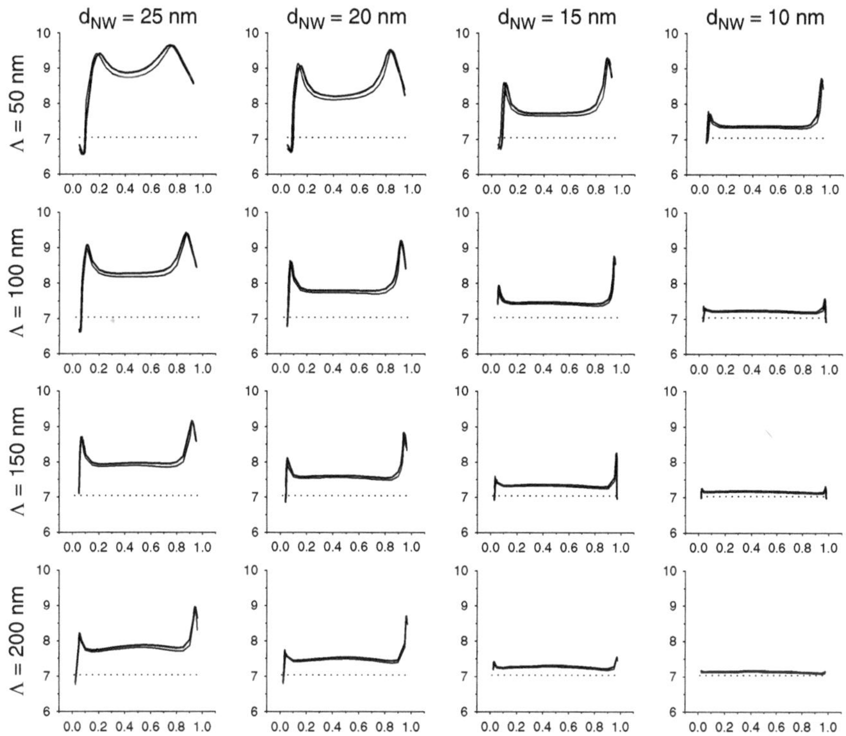

Fig. 5 Plasmon momentum k_{SP} (y-axis in unit of μm^{-1}) for various combinations of nanowire depths (d_{NW}) and periods (Λ). x-axis is VF. Thick and thin solid lines represent the characteristics of a nanowire-based SPR biosensor with and without a HDT SAM as target analytes. For comparison, the dotted line shows $k_{SP}(\theta_{SP} = 45.13°)$ of a SPR biosensor at $d_f = 40$ nm without nanowires. Reprinted from [18], Copyright (2006), with permission from the Optical Society of America

decoupled LSP mode, compared with the one coupled to a bulk SP mode, has been numerically investigated by Byun et al. [21].

Second, the LSPR is associated with two different modes of enhancement at a small or a large value of VF. For convenience, I define the VF at maximum LSPR enhancement as VF_L and VF_H, respectively for the enhancement associated with single-nanowire LSP excitation (low VF) and that associated with the resonant coupling of LSPs through a nanogroove (high VF). Note that the LSP excitation associated with a nanogroove tends to be noticeably stronger than that of a single nanowire. The data in Fig. 5 also suggest that the two LSP excitation modes be coupled with effective LSPR enhancement. In other words, LSPR enhancement is reduced if the coupling between the two modes is weakened for a large nanowire period. At a very small period and large depth (see the case of $\Lambda = 50$ nm and $d_{NW} = 25$ nm), it is apparent that the two modes of enhancement begin to merge and create significant coupling of LSPs excited in neighboring nanowires.

Third, in the presence of a SAM that models the target binding, VF_L becomes smaller and VF_H larger. As in Fig. 5, the presence of a SAM not only shifts the resonance condition but also causes the resonance to occur more easily with a SAM, i.e.,

$$VF_L(\text{SAM}) > VF_L(\text{no SAM}) \text{ and } VF_H(\text{SAM}) < VF_H(\text{no SAM}) \qquad (13)$$

Comparison of the resonance at $\Lambda = 100$ nm and $d_{NW} = 20$ nm with the one at $\Lambda = 50$ nm and $d_{NW} = 10$ nm implies that the impact of depth (d_{NW}) may be more significant than that of nanowire period (Λ), i.e., deeper nanowires yield larger LSPR enhancement at an identical ratio of Λ/d_{NW}. This is supported by comparing the resonance of $\Lambda = 100$ nm and $d_{NW} = 10$ nm versus $\Lambda = 200$ nm and $d_{NW} = 20$ nm. This can be a basis of implementing a highly sensitive LSPR biosensor, since it hints at a possibility of sensitivity enhancement using relatively coarse nanowires so that they can be fabricated based on conventional lithography and simply by adjusting the depth for sensitivity control, a far easier parameter to manipulate than the nanowire period.

2.1.4.4 Profile

Nanowire profile also affects the excitation behavior of LSPs, i.e., it is one of the critical parameters that determine how electrons redistribute themselves in order to meet momentum matching condition. Effects of many profiles, such as rectangular, triangular, T and inverse T profiles, have been investigated [22, 23].

Intuitively, a T profile produces more localized fields than an inverse T profile, likewise an inverse trapezoidal profile does than a trapezoidal profile. This trend has been confirmed in the earlier reports. Consistent with the previous discussion on the effect of depth, higher localization seems to be correlated with isolation of LSP modes from propagating bulk SPs to some degree. The field enhancement near vertices has been explained by [23] as "corner effects," which should be differentiated from edge effect of electromagnetic fields.

2.1.5 Application Issues: Linearity

An LSPR biosensor measures refractive index changes accompanied by a biomolecular interaction. An index change as an interaction progresses is not always linear. In practice, the linearity between index changes and resonance condition is more a concern for a biosensor. Plasmon signals represented as a change in resonance angles or wavelengths are fairly linear with an index change. However, when a signal is strong, nonlinearity prevails such that an increased signal does not create a shift in resonance condition or even brings a negative shift. This is due to the intrinsic nonlinearity in the dispersion relation of (1) and thus appears in a conventional thin film-based SPR structure as well, as shown in Fig. 6 [24]. The nonlinearity

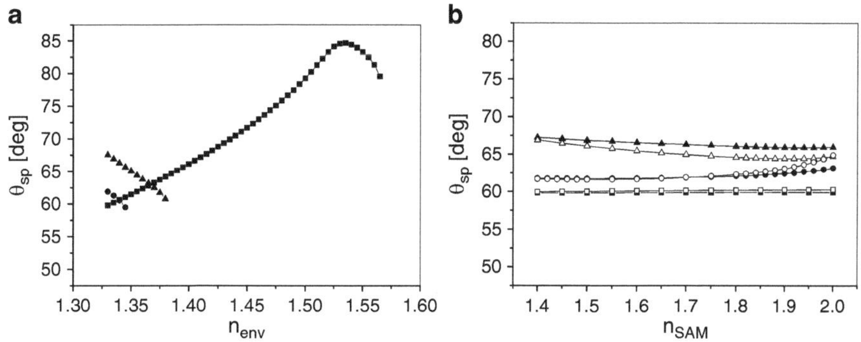

Fig. 6 Resonance angles on an SF10 substrate with a refractive index of (**a**) a bulk environment n_{env} (■: conventional SPR, ●: LSPR with $\Lambda = 50$ nm, and ▲: LSPR with $\Lambda = 100$ nm) and (**b**) a target binding layer n_{SAM}. (filled: $d_{SAM} = 1$ nm and empty: $d_{SAM} = 3$ nm) Reprinted from [24], Copyright (2008), with permission from the Optical Society of America

was also reported by Jung et al. [25]. Because stronger signals are more likely to produce nonlinear resonance shifts, an LSPR biosensor is inflicted more with the nonlinearity. This may restrict the dynamic range of an LSPR biosensor relative to the conventional thin film-based SPR structure.

2.1.6 Experimental Results

LSPR-based sensitivity enhancement using surface-relief nanostructures has been confirmed experimentally in a few studies to date. In the experiments conducted by Byun et al. [26], ethanol–water mixture at varied ethanol concentration was used to estimate the sensitivity enhancement by periodic nanowires at $\Lambda = 200$ nm and 500 nm respectively as 44% and 31% over conventional structures, as shown in Fig. 7. Note that the sensitivity enhancement for bulk index measurement is relatively limited compared to layered bio molecular interactions, because of reduced index contrast against ambience. It was also found that surface roughness can degrade sensitivity performance [27]. Measurement of the DNA hybridization process was performed using nanoposts at $\Lambda = 110$ nm and presented more than fivefold sensitivity improvement, as shown in Fig. 8 [28].

2.2 *LSPR Bio-sensor Based on Coupling Between Particle Plasmon and LSP Excited by Surface Relief Nanostructures*

Here, I examine the coupling of particle plasmons excited in nanoparticles with LSPs in surface relief nanostructures. As a biosensor, nanoparticles may serve as linker molecules that amplify the index change due to ligand bindings with

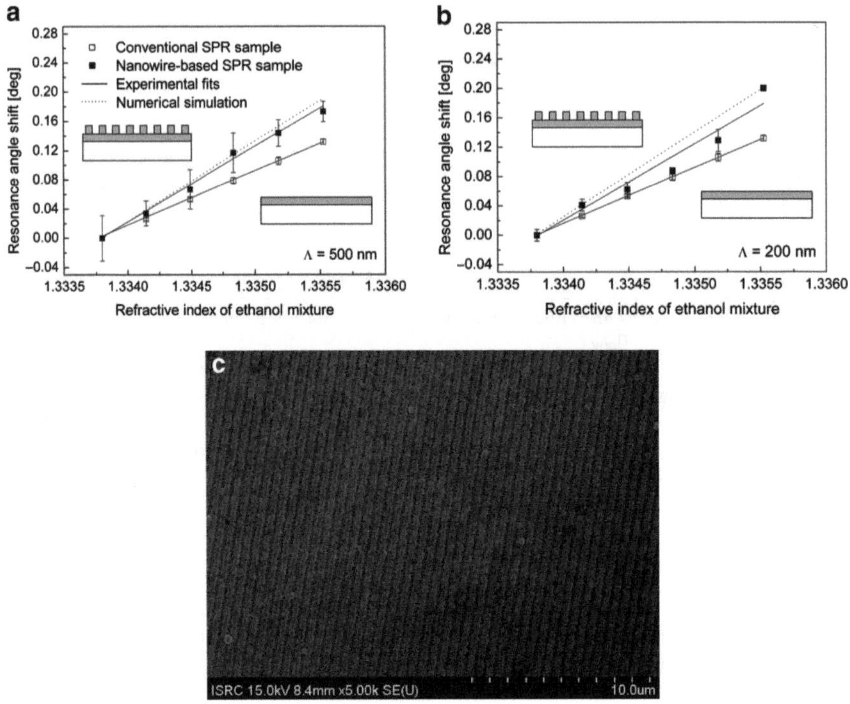

Fig. 7 Resonance angle change for ethanol-water mixture at varied ethanol concentration. Measurements using periodic nanowires at (**a**) $\Lambda = 200$ and (**b**) 500 nm show sensitivity enhancement by 44% and 31% over conventional SPR structures. A SEM image of the nanowire sample is also shown in (**c**). Reprinted from [26], Copyright (2007), with permission from the Optical Society of America

surface-coated antibodies. At times, nanoparticles themselves can be the target of interest to be measured. The amplification can be associated with two mechanisms. First, the presence of metal nanoparticles involves much larger index changes than the binding without nanoparticles. Moreover, LSP coupling may bring significant local field enhancement that provides additional sensitivity to an index change.

2.2.1 Geometry

The geometry of interest can be simplified as in Fig. 9. Nanoparticles are functionalized with ligands for attachment to surface-relief nanostructures While Fig. 9 omits details of surface chemistry, such as a linker layer, the structure implicates that the previous analysis remains valid overall.

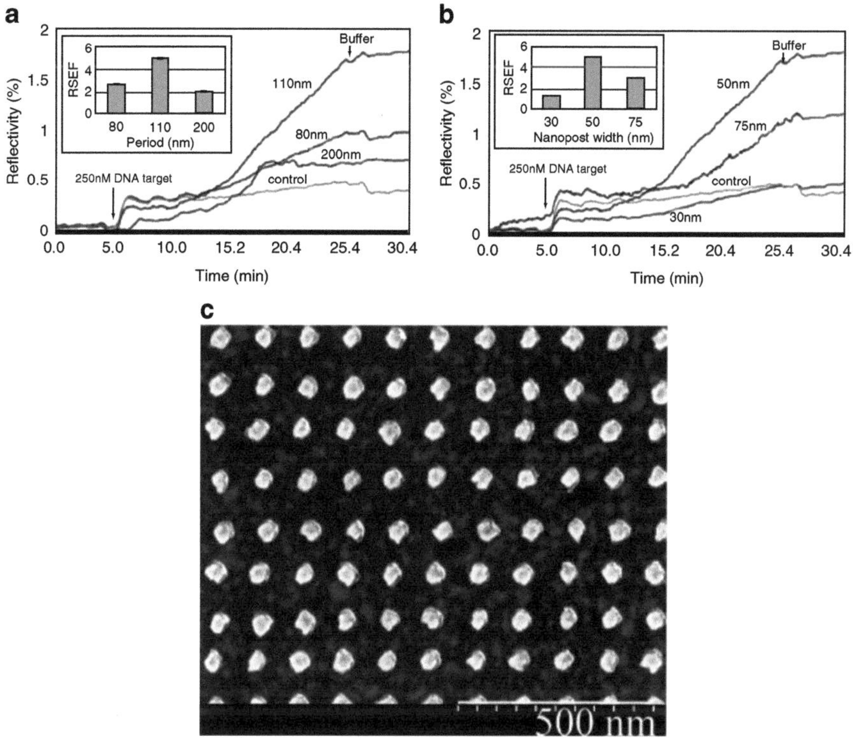

Fig. 8 Reflectivity change at resonance of 250 nM DNA hybridization on an LSPR substrate with nanoposts that are 30 nm in height on a 20 nm thick underlying gold film with (**a**) 50 nm width of various periods and (**b**) 110 nm period of various widths. (**c**) An SEM image of fabricated nanoposts with 50 nm width and 110 nm period. Reprinted from [28], Copyright (2007), with permission from the Optical Society of America

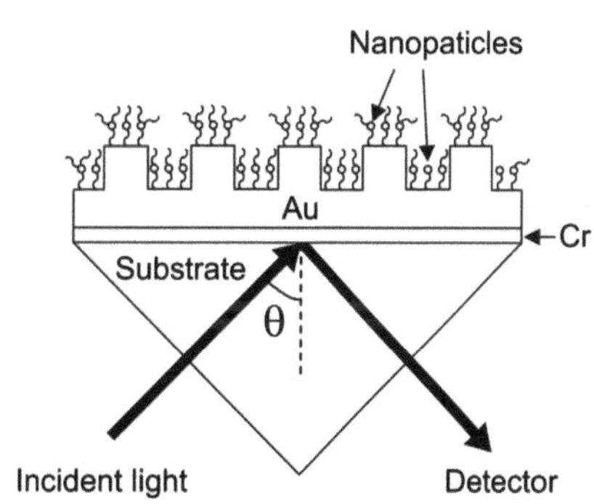

Fig. 9 Schematic of an LSPR bio-sensor based on periodic nanowires coupled to particle plasmons

2.2.2 Dispersion Relation

An effective index for nanoparticles can be obtained, for example, using the Maxwell Garnett EMT [29, 30] such that

$$\frac{\varepsilon_{\mathrm{np}} - \varepsilon_{\mathrm{A}}}{\varepsilon_{\mathrm{np}} + 2\varepsilon_{\mathrm{A}}} = f\frac{\varepsilon_{\mathrm{B}} - \varepsilon_{\mathrm{A}}}{\varepsilon_{\mathrm{B}} + 2\varepsilon_{\mathrm{A}}}, \tag{14}$$

where $\varepsilon_{\mathrm{np}}$ is an effective permittivity of nanoparticles without surface-relief nanostructures. The total effective permittivity $\varepsilon_{\mathrm{eff}}$ can be modeled simply as a combination of $\varepsilon_{\mathrm{np}}$ and that of target layers that include antibody-target binding. By inserting the effective permittivity for the combined target layers into (4) or (5), the dispersion relation can be obtained. The procedure becomes complicated in the case that particle plasmons excited in nanoparticles are coupled to LSPs in surface-relief nanostructures, mainly because a simple EMT cannot be directly applied to describing the case. Assuming that the perturbation due to the surface-relief nanostructure is not significant, the dispersion relation can be estimated qualitatively in a manner similar to the discussion presented in (2.1.2).

2.2.3 Effects of Particle Parameters

Not many studies have experimentally addressed the interaction between particle plasmons and LSPs created by surface nanostructures, let alone the application of the interaction from biological engineering perspectives. In contrast, there are quite a few studies that investigate particle plasmons in plasmonic fields created by a conventional thin film-based SPR configuration, oftentimes using angle-resolved techniques. Obviously, use of nanoparticles amplifies plasmon signals, which was experimentally confirmed by measuring a reaction of human IgG conjugated with gold colloids of 10 nm diameter, as shown in Fig. 10 [31]. Amplification of DNA

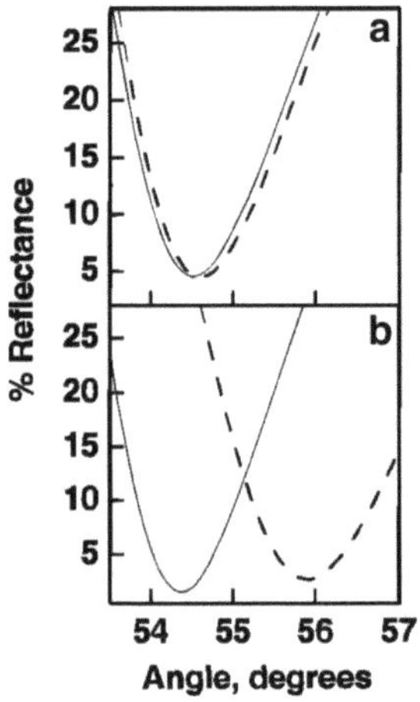

Fig. 10 In situ SPR curves of (**a**) an evaporated Au film modified with a-h-IgG() (-) and then exposed to a 1.0 mg/mL solution of h-IgG (- -). (**b**) A film modified with a-h-IgG() (-) and then exposed to h-IgG-10-nm Au colloid conjugate (- -)]. Reproduced with permission from [31]. Copyright 1998 American Chemical Society

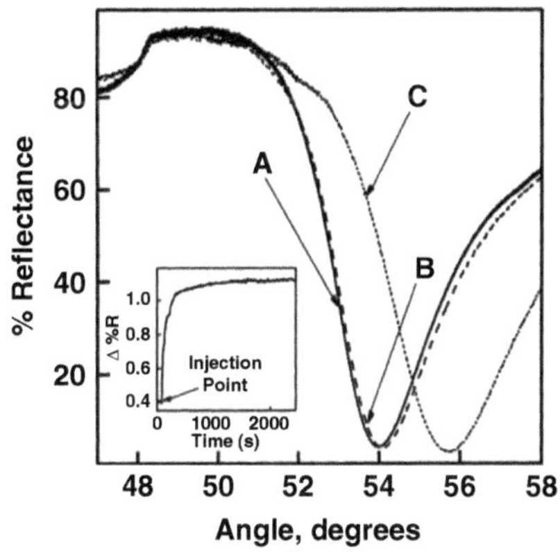

Fig. 11 SPR curves of surfaces prepared in sequential steps as illustrated in Scheme 1: a MHA-coated Au film modified with a 12-mer oligonucleotide S1(**a**), after hybridization with its complementary 24-mer target S2 (**b**), and followed by introduction of S3:Au conjugate (**c**) to the surface. Inset: surface plasmon reflectance changes at 53.2° for the oligonucleotide-coated Au film measured during a 60-min exposure to S3:Au conjugates. Reproduced with permission from [32]. Copyright 2000 American Chemical Society

hybridization processes was observed when surface-coated DNA oligomers were bound to the complementary strands that were conjugated with gold nanoparticles of 12 nm diameter (shown in Fig. 11) [32].

In the case of using a substrate with surface-relief nanostructures, overall trends would agree with what is mentioned in (Sect. 2.1.2), i.e., momentum-matching would occur at higher plasmon momentum. Since gold nanoparticles act as a target of supermolecules, plasmon momentum would be shifted further, which may induce nonlinear plasmon characteristics. Here, the effects of particle parameters, such as particle size, concentration, shape, and interaction distance between metal surface and nanoparticles, are discussed briefly based on experimental data of the interaction between particle plasmons and conventional SPs.

2.2.3.1 Size

To the first degree, larger nanoparticles exhibit stronger plasmon signatures. Note that a gold nanoparticle of 10 nm diameter weighs approximately 6 MDa assuming that its density is equal to that of bulk at 19.3 g/cm^3. Experimental confirmation is presented in Fig. 12 that shows SPR curves of a submonolayer of colloidal gold particles coated with 2-mercaptoethylamine on a 47-nm-thick gold film [33]. The angular shift is marginal between SPR curves of nanoparticles of 45 nm diameter and those of 59 nm diameter. This suggests an onset of nonlinear sensitivity for strong SPR signals.

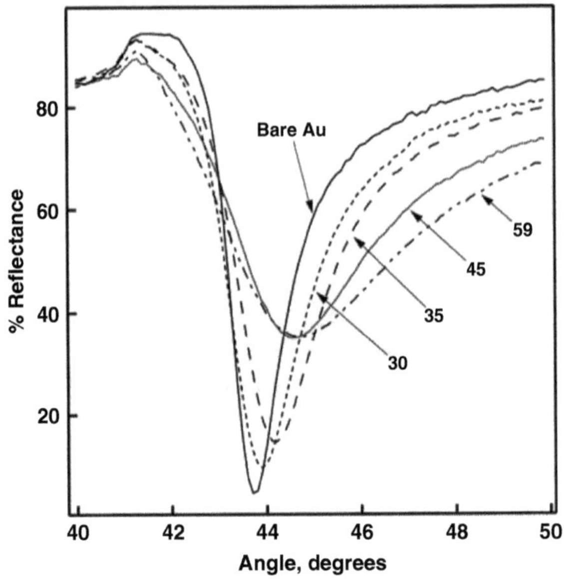

Fig. 12 SPR curves (reflectance vs angle) of a 47 nm thick Au film coated with 2-mercaptoethylamine (-) and a submonolayer of colloidal Au ($1.3 \pm 0.15 \times 10^9$ particles/cm^2). Particle diameters are 30 (*dotted*), 35 (*dashed*), 45 (*solid*), and 59 nm (*double dot dashed*). Reproduced with permission from [33]. Copyright 1999 American Chemical Society

2.2.3.2 Concentration

Increased nanoparticle concentration is presumed to encourage stronger plasmon signals, because more particle plasmons can be excited. The plasmon characteristics are not linear with nanoparticle concentration since the characteristics depend on the plasmonic fields created by surface-relief nanostructures. In other words, the relationship between plasmon signals and concentration is affected by many parameters such as the relative dimension and geometry of nanoparticles and surface nanostructures. Experimental SPR curves and plasmon angle shifts are shown in Figs. 13 and 14, respectively, for a submonolayer of colloidal gold particles coated with 2-mercaptoethylamine on a 47-nm-thick gold film [33].

2.2.3.3 Shape

Nanoparticles of different shape affect plasmon characteristics, since the shape changes the way particle plasmons are excited. Overall trends of the plasmon shift are consistent with Sect. 2.1.4.4. For example, stronger plasmons tend to be induced in nanoparticles with corners. In other words, isolated LSPs may produce a more significant shift.

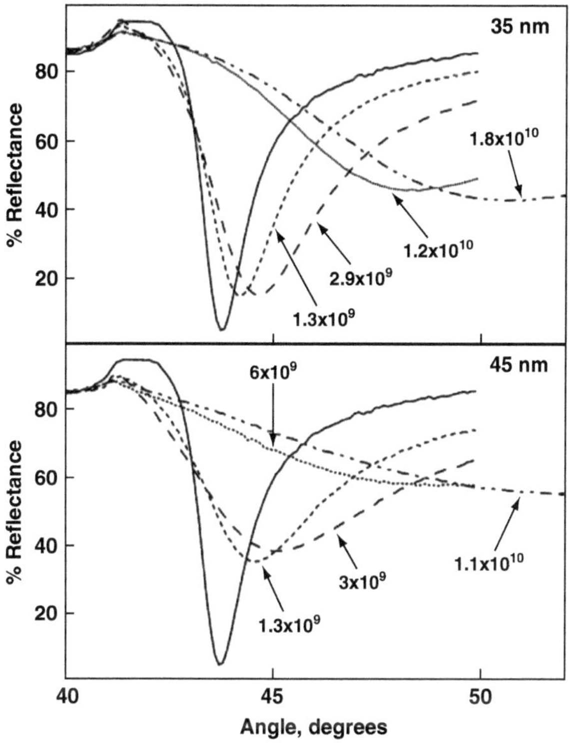

Fig. 13 SPR curves (reflectance vs angle) of a 47 nm thick Au film coated with 2-mercaptoethylamine (-) and a layer of 35 nm diameter colloidal Au (*top*) and 45 nm diameter colloidal Au (*bottom*). Colloid surface concentrations are in particles/cm^2. Reproduced with permission from [33]. Copyright 1999 American Chemical Society

2.2.3.4 Interaction Distance

Physically, the interaction distance refers to the separation between nanoparticles and surface-relief nanostructures. At the long interaction distance, exponentially decreasing evanescent waves created by LSPs in surface-relief nanostructures excite very weak particle plasmons in nanoparticles, thus little coupling between particle plasmons and LSPs in surface relief nanostructures. Metallic quenching can arise at a close proximity within 10 nm and can extinguish the coupling effectively [34, 35], although the degree of quenching may differ depending on specific particle parameters. This suggests that an optimum distance should exist, which induces maximal plasmon coupling. Maximal plasmon shift at an optimum distance was reported by Holland and Hall [36], as shown in Fig. 15, using a silver particulate island and silver thin film with the separation defined by a LiF film. Changes in the dispersion relation and resonance broadening as a result were studied and visualized by Kume et al. [37], as shown in Fig. 16.

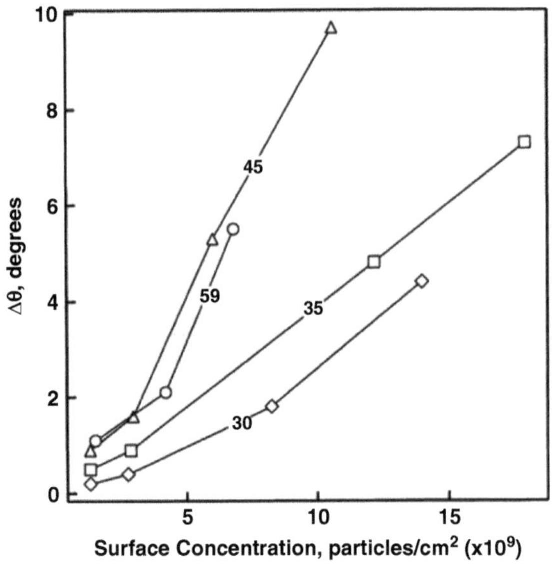

Fig. 14 The plasmon angle shift plotted as a function of particle number density for 30 (open diamond), 35 (open rectangle), 45 (open triangle), and 59 nm (open circle) diameter colloidal Au. Reproduced with permission from [33]. Copyright 1999 American Chemical Society

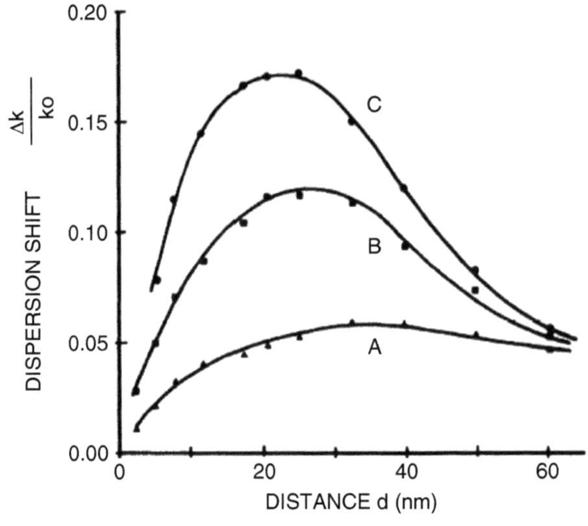

Fig. 15 Resonance shift Δk with spacer layer thickness d for $\lambda = 632.8$ nm (*curve A*), 514.5 nm (*curve B*), and 476.5 nm (*curve C*). Δk represents the difference between the measured propagation constant k for the Ag–LiF–Ag islands-air and the Ag–LiF–air cases. Reprinted figure with permission from [36]. Copyright (1983) by the American Physical Society

2.3 LSPR Bio-sensor Based on Particle Plasmons

An LSPR bio-sensor based on particle plasmons uses plasmon excitation in small metal particles that are often sized at a few tens of nanometers. Metallic nanoparticles

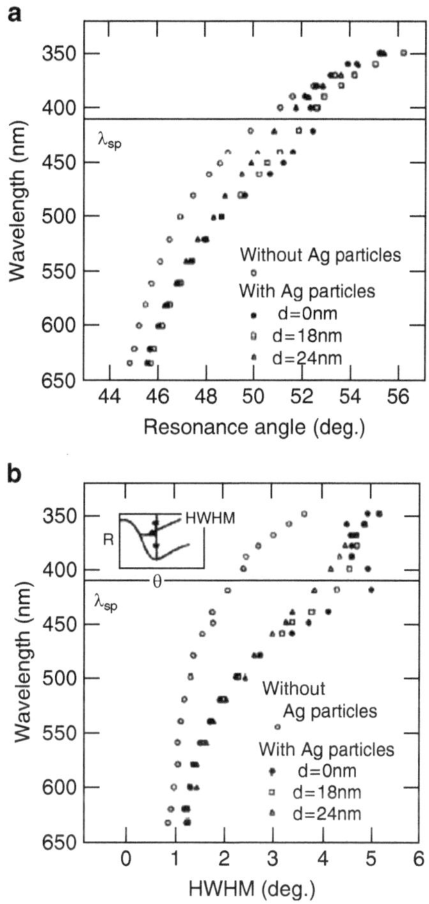

Fig. 16 (**a**) Dispersion relation (w vs. k) (**b**) Wavelength dependence of the half width at half minimum. Reprinted from [37], Copyright (1995), with permission from Elsevier

are extremely flexible since they can be functionalized with linker molecules, ligands, and antibodies. By the nature of particle plasmons, this type of LSPR biosensors scans wavelengths in SPR spectroscopy and measures the changes in transmittance or extinction during an interaction. The measurement schematic would be close to Fig. 17 where functionalized nanoparticles are suspended in solution. If metallic nanoparticles conjugated with specific antibodies in solution bind to specific target molecules, the binding interaction would change plasmon resonance condition. The change appears as a shift in the absorption wavelength. However, no resonance wavelength shift indicates that the conjugated metallic particles have not bound to the molecules. Figure 18 presents an example of biosensing using LSPR based on biotinylated silver nanoparticles which shows the change of resonance patterns in the course of surface modification processes [38].

In the case of binding, the resonance wavelength shift arises, which relies on many factors, for example, whether target molecules are metallic or dielectric and

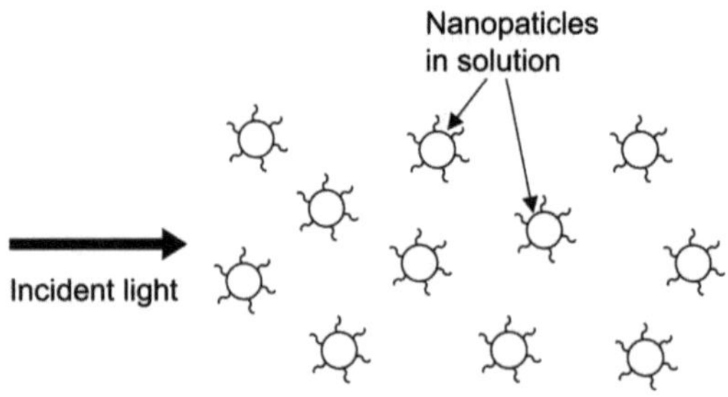

Fig. 17 Schematic of an LSPR bio-sensor based on particle plasmons

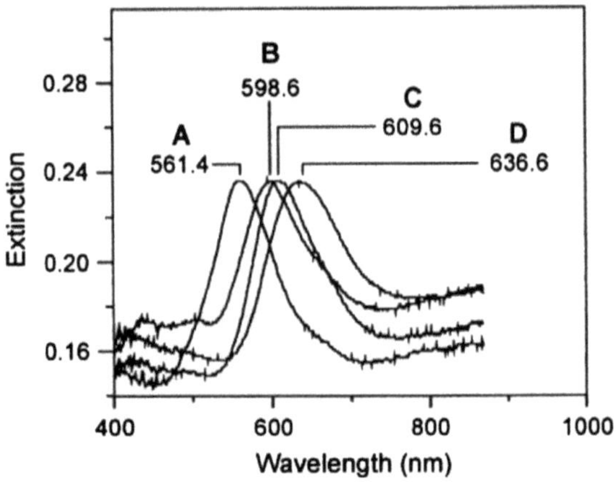

Fig. 18 LSPR spectra of each step in the surface modification of Ag nanoparticles to form a biotinylated Ag nanobiosensor and the specific binding of streptavidin. (**a**) Ag nanoparticles before chemical modification, $\lambda_{max} = 561.4$ nm. (**b**) Ag nanoparticles after modification with 1 mM 1:3 11-MUA/1-OT, $\lambda_{max} = 598.6$ nm. (**c**) Ag nanoparticles after modification with 1 mM biotin, $\lambda_{max} = 609.6$ nm. (**d**) Ag nanoparticles after modification with 100 nM streptavidin, $\lambda_{max} = 636.6$ nm. Reproduced with permission from [38]. Copyright 2002 American Chemical Society

volume, weight, and concentration of nanoparticles and target molecules. Note that there is little coupling between plasmonic fields induced in these particles. Before significant coupling is induced, nanoparticles tend to aggregate and the aggregation affects the resonance wavelength significantly.

An LSPR bio-sensor based on particle plasmons should, however, be distinguished from the surface mediated bio-sensors that were described earlier. First of all, its plasmonic characteristics are represented by scattering properties of metallic particles and can be described precisely using Rayleigh or Mie scattering theory. In most cases, plasmon characteristics appear for both p and s polarization. Intrinsically, the plasmon characteristics are not polarization sensitive, as long as nanoparticles or nanorods are randomly suspended in solution. Also, angle-scanning does not characterize plasmon shifts of an LSPR biosensor based on particle plasmons, other than varied light interaction lengths through nanoparticle

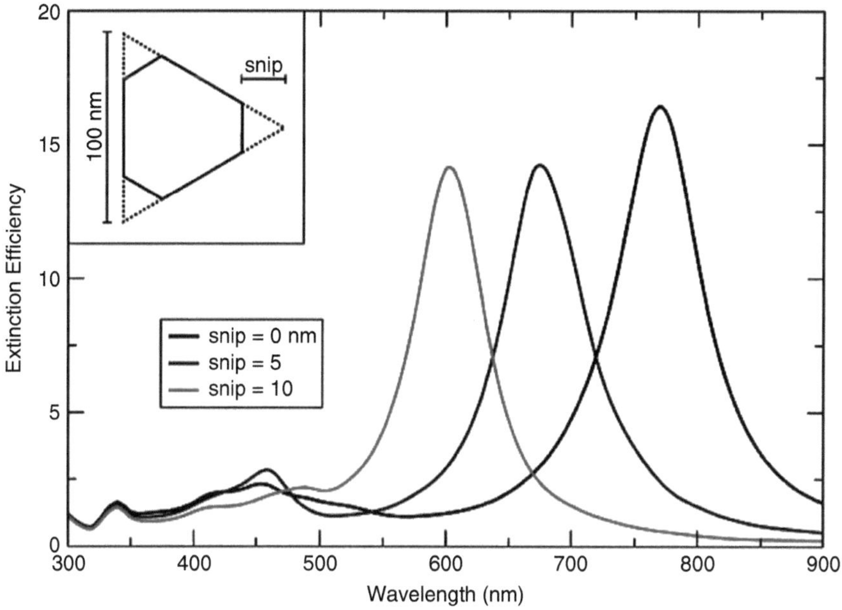

Fig. 19 Calculated orientation-averaged extinction efficiency for trigonal prisms based on a 100 nm edge dimension with snips of 0, 10, and 20 nm (right to left). The inset shows the shape of a snipped prism. The prism thickness is 16 nm. Reproduced with permission from [39]. Copyright 2003 American Chemical Society

solution and refractive changes at sample–ambience interfaces. Thus, spectroscopic methods are employed almost exclusively for characterizing particle plasmons.

Numerous reports have been published on the plasmon characteristics of various types of nanoparticles. While I do not attempt to recapitulate or summarize the results here, the general trends of plasmon characteristics discussed previously remain equally valid to single particle plasmons. For example, Fig. 19 describes plasmon characteristics of nanoparticles as blue-shifted with more vertices [39]. The shape dependence is consistent with particle plasmon characteristics measured by LSPR spectroscopy on pyramidal surface structures or surface-immobilized nanoparticles. The shape dependence has been described in (2.1.4.4) and (2.2.3.3) in terms of corner effects, in that higher plasmon momentum is associated with structures that have more vertices, as shown in Fig. 20 [40]. Therefore, rod-like nanoparticles exhibit higher sensitivity (triangular ones the next) than spherical nanoparticles, as shown in Fig. 21 [41]. More significant enhancement using silver nanoparticles than gold has also been reported consistently (see Fig. 22 [40]). Figure 22 also exhibits the nonlinearity of plasmon signals with respect to nanoparticle geometry. However, target-dependent plasmon shift of particle LSPR was measured by varying alkanethiol chain lengths conjugated to silver nanoparticles on a glass substrate, as shown in Fig. 23 [42].

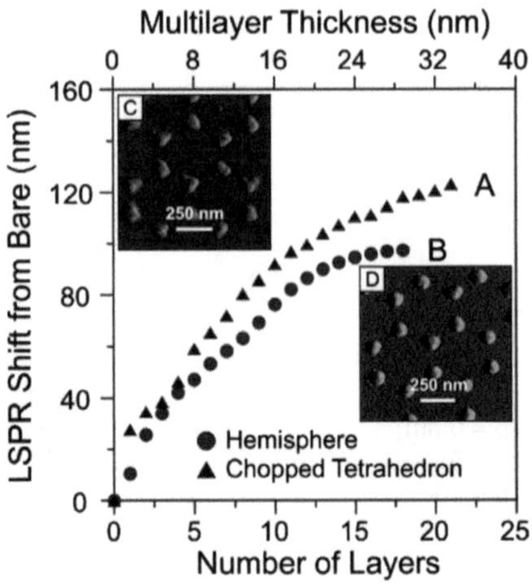

Fig. 20 Shape dependence on the long range LSPR distance dependence for Ag nanoparticles ($D = 400$ nm). (**a**) LSPR shift vs number of SAM layer thickness for solvent annealed Ag nanoparticles. (**b**) LSPR shift vs number of SAM layer thickness for thermally annealed (600 C for 1 h) Ag nanoparticles. (**c**) AFM image of solvent annealed nanoparticles ($a = 114$ nm, $b = 54$ nm). (**d**) AFM image of thermally annealed nanoparticles ($a = 110$ nm, $b = 61$ nm). Reproduced with permission from [40]. Copyright 2004 American Chemical Society

Fig. 21 Comparison of refractive index sensitivity for Ag nanoparticles with different geometries. The spherical nanoparticle (●) has a sensitivity of 161 nm per RIU, the triangular nanoparticle (▲) has a sensitivity of 197 nm per RIU, and the rodlike nanoparticle (■) has a sensitivity of 235 nm per RIU. Reproduced with permission from [41]. Copyright 2003 American Chemical Society

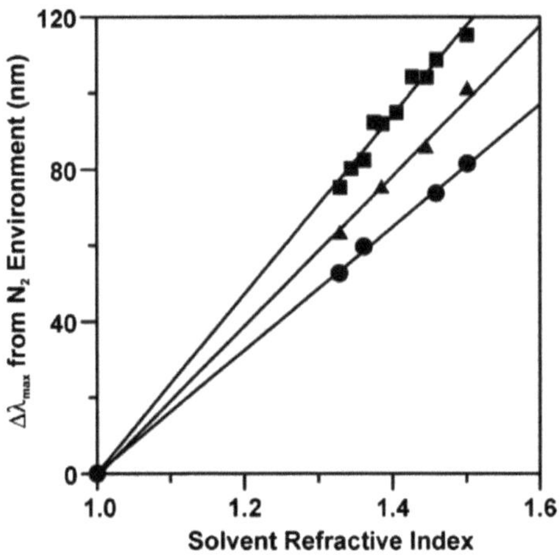

It is also important to notice that plasmon characteristics of particle plasmons typically represent those of single particles, because particles tend to aggregate at high concentration at which interparticle interaction starts to appear for an ensemble of particles and particle plasmon resonance is affected additionally by particle

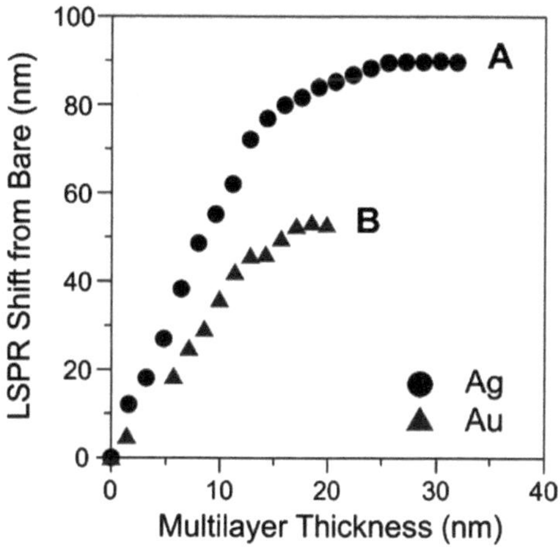

Fig. 22 Composition dependence on the long range LSPR distance dependence. (**a**) LSPR shift vs layer thickness for Ag nanoparticles ($a = 70$ nm, $b = 50.0$ nm). (**b**) LSPR shift vs layer thickness for Au nanoparticles ($a = 70$ nm, $b = 50.0$ nm). Multilayer thicknesses were adjusted for SAM packing differences on Au and Ag. Reproduced with permission from [40]. Copyright 2004 American Chemical Society

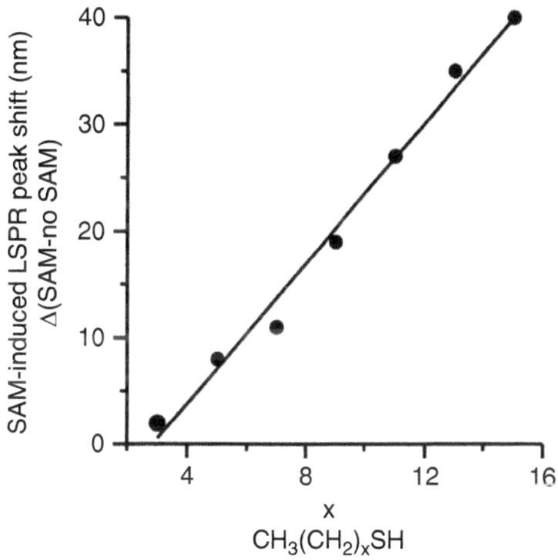

Fig. 23 Alkanethiol chain length dependence on the LSPR spectral peak shift. All extinction measurements were collected in a N_2 environment. Linear regression was used to fit the data to a line described by the following equation: $y = 3.3(x) - 9.3$. Reproduced with permission from [42]. Copyright 2001 American Chemical Society

interaction. The influence of the particle interaction on the plasmon resonance was the subject of [43].

3 Concluding Remarks

Nanostructure-based LSPR biosensing has been classified into three schemes: surface-relief nanostructures, surface-relief nanostructures coupled to nanoparticles, and nanoparticles. These schemes share many aspects of plasmon characteristics in common, such as shape and concentration dependence. Local field enhancement as a result of plasmon excitation can be used for highly sensitive biosensing. Given particular bio-sensing applications, one of these schemes can be selected to meet specific requirements, e.g., need for functionalization, spatial localization of target molecules, and detection sensitivity.

References

1. Strutt JW (Lord Rayleigh) (1907) Dynamical theory of the grating. Proc Roy Soc A79: 399–416
2. Wood RW (1902) On a remarkable case of uneven distribution of light in a diffraction grating spectrum. Philos Magazine 4:396–402
3. Liedberg B, Nylanderm C, Lundström I (1983) Surface plasmon resonance for gas detection and biosensing. Sensors Actuators 4:299–304
4. Liedberg B, Nylanderm C, Lundström I (1995) Biosensing with surface plasmon resonance – how it all started. Biosens Bioelectron 10:1–9
5. Malmqvist M (1993) Biospecific interaction analysis using biosensor technology. Nature 361:186–187
6. Abdulhalim I, Zourob M, Lakhtakia A (2008) Surface plasmon resonance for biosensing: a mini-review. Electromagnetics 28:214–242
7. Salomon Z, Lindblom G, Rilfors L et al (2000) Interaction of phosphatidylserine synthase from E. coli with lipid bilayers: coupled plasmon-waveguide resonance spectroscopy studies. Biophys J 78:1400–1412
8. Raether H (1988) Surface plasmon on smooth and rough surfaces and on gratings. Springer, Berlin, Germany
9. Campbell CT, Kim G (2007) SPR microscopy and its applications to high-throughput analyses of biomolecular binding events and their kinetics. Biomaterials 28:2380–2392
10. Ran B, Lipson SG (2006) Comparison between sensitivities of phase and intensity detection in surface plasmon resonance. Opt Express 14:5641–5650
11. Rytov SM (1956) Electromagnetic properties of a finely stratified medium. Sov Phys JETP 2:466–475
12. Kim D, Yoon SJ (2007) Effective medium-based analysis of nanowire-mediated localized surface plasmon resonance. Appl Opt 46:872–880
13. Sherry LJ, Chang SH, Schatz GC et al (2005) Localized surface plasmon resonance spectroscopy of single silver nanocubes. Nano Lett 5:2034–2038
14. Kim D (2006) Effect of resonant localized plasmon coupling on the sensitivity enhancement of nanowire-based surface plasmon resonance biosensors. J Opt Soc Am A 23:2307–2314
15. Ebbesen TW, Lezec HJ, Ghaemi HF et al (1998) Extraordinary optical transmission through sub-wavelength hole arrays. Nature 391:667–669
16. Genet C, Ebbesen TW (2007) Light in tiny holes. Nature 445:39–46
17. Yoon SJ, Kim D (2007) Thin-film-based field penetration engineering for surface plasmon resonance biosensing. J Opt Soc Am A 24:2543–2549
18. Kim K, Yoon SJ, Kim D (2006) Nanowire-based enhancement of localized surface plasmon resonance for highly sensitive detection: a theoretical study. Opt Express 14:12419–12431
19. Pockrand I, Raether H (1976) Surface plasma oscillations in silver films with wavy surface profiles: a quantitative experimental study. Opt Commun 18:395–399
20. Raether H (1982) Dispersion relation of surface plasmons on gold- and silver gratings. Opt Commun 42:217–222
21. Byun KM, Kim SJ, Kim D (2006) Profile effect on the feasibility of extinction-based localized surface plasmon resonance biosensors with metallic nanowires. Appl Opt 45:3382–3389
22. Byun KM, Kim SJ, Kim D (2005) Design study of highly sensitive nanowire-enhanced surface plasmon resonance biosensors using rigorous coupled wave analysis. Opt Express 13:3737–3742

23. Kottmann JP, Martin OJF, Smith DR et al (2001) Plasmon resonances of silver nanowires with a nonregular cross section. Phys Rev B 64:235402
24. Yoon SJ, Kim D (2008) Target dependence of the sensitivity in periodic nanowire-based localized surface plasmon resonance biosensors. J Opt Soc Am A 25:725–735
25. Jung LS, Campbell CT, Chinowsky TM et al (1998) Quantitative interpretation of the response of surface plasmon resonance sensors to adsorbed films. Langmuir 14:5636–5648
26. Byun KM, Yoon SJ, Kim D et al (2007) Experimental study of sensitivity enhancement in surface plasmon resonance biosensors by use of periodic metallic nanowires. Opt Lett 32:1902–1904
27. Byun KM, Yoon SJ, Kim D et al (2007) Sensitivity analysis of a nanowire-based surface plasmon resonance biosensor in the presence of surface roughness. J Opt Soc Am A 24: 522–529
28. Malic L, Cui B, Veres T et al (2007) Enhanced surface plasmon resonance imaging detection of DNA hybridization on periodic gold nanoposts. Opt Lett 32:3092–3094
29. Maxwell Garnett JC (1904) Colours in metal glasses and in metallic films. Philos Trans R Soc A 203:385–420
30. Maxwell Garnett JC (1906) Colours in metal glasses, in metallic films and in solutions II. Philos Trans R Soc A 205:237–288
31. Lyon LA, Musick MD, Natan MJ (1998) Colloidal Au-enhanced surface plasmon resonance immunosensing. Anal Chem 70:5177–5183
32. He L, Musick MD, Nicewarner SR et al (2000) Colloidal Au-enhanced surface plasmon resonance for ultrasensitive detection of DNA hybridization. J Am Chem Soc 122:9071–9077
33. Lyon LA, Pena DJ, Natan MJ (1999) Surface plasmon resonance of Au colloid-modified Au films: particle size dependence. J Phys Chem B 103:5826–5831
34. Daffertshofer M, Port H, Wolf HC (1995) Fluorescence quenching of ultrathin anthracene films by dielectric and metallic substrates. Chem Phys 200:225–233
35. Yu F, Persson B, Löfås S et al (2004) Attomolar sensitivity in bioassays based on surface plasmon fluorescence spectroscopy. J Am Chem Soc 126:8902–8903
36. Holland WR, Hall DG (1983) Surface-plasmon dispersion relation: shifts induced by the interaction with localized plasma resonances. Phys Rev B 27:7765–7768
37. Kume T, Nakagawa N, Hayashi S et al (1995) Interaction between localized and propagating surface plasmons: Ag fine paticles on Al surface. Solid State Commun 93:171–175
38. Haes AJ, Van Duyne RP (2002) A nanoscale optical biosensor: sensitivity and selectivity of an approach based on the localized surface plasmon resonance spectroscopy of triangular silver nanoparticles. J Am Chem Soc 124:10596–10604
39. Kelly KL, Coronado E, Zhao LL et al (2003) The optical properties of metal nanoparticles: the influence of size, shape, and dielectric environment. J Phys Chem B 107:668–677
40. Haes AJ, Zou S, Schatz GC et al (2004) A nanoscale optical biosensor: the long range distance dependence of the localized surface plasmon resonance of noble metal nanoparticles. J Phys Chem B 108:109–116
41. McFarland AD, Van Duyne RP (2003) Single silver nanoparticles as real-time optical sensors with zeptomole sensitivity. Nano Lett 3:1057–1062
42. Malinsky MD, Kelly KL, Schatz GC et al (2001) Chain length dependence and sensing capabilities of the localized surface plasmon resonance of silver nanoparticles chemically modified with alkanethiol self-assembled monolayers. J Am Chem Soc 123:1471–1482
43. Lamprecht B, Schider G, Lechner RT et al (2000) Metal nanoparticle gratings: influence of dipolar particle interaction on the particle resonance. Phys Rev Lett 84:4721–4724

Gold Nanoparticles on Waveguides For and Toward Sensing Application

Silvia Mittler

Abstract First, a short overview of the sensor activities on surface plasmon waveguide mode coupling via wave vector match in metal-coated channel waveguides as well as in slab waveguides and optical fibers is given. Both monomode and multimode approaches were demonstrated as well as the implementation of Bragg gratings and hollow fibers are described. Then, the use of gold nanoparticles for sensor application in combination with these optical device systems is discussed. At the beginning, a channel waveguide approach with gold nanoparticles is described without taking the typical optical features of gold nanoparticles into account. Then, waveguide devices, which use the localized surface plasmon resonance, an absorption band, intrinsic to gold nanoparticles, and the color changes of gold colloids upon clustering for sensor operation are described. This is achieved on quasi-waveguides, channel waveguides, and on optical fibers. Transmission and reflectance experiments have been performed, either with spectral information or with monochromatic light and pure intensity information. An electro-optical approach is discussed. The activities in photonic crystal sensor are described for planar-waveguide systems and hollow photonic crystal fiber bundles.

Keywords Optical waveguides · Optical fibers · Thin gold layers · Gold nanoparticles · Photonic crystals · Sensors

Contents

S. Mittler
Department of Physics and Astronomy, University of Western Ontario, London, Ontario N6A 3K7, Canada
e-mail: smittler@uwo.ca

M. Zourab and A. Lakhtakia (eds.), *Optical Guided-wave Chemical and Biosensors I*, Springer Series on Chemical Sensors and Biosensors 7,
DOI 10.1007/978-3-540-88242-8_8,

Abbreviations

APTES	3- aminopropyltriethoxysilane
Au NPs	Gold nanoparticles
CCD	Charged coupled device
DDA	Discrete dipole approximation
DNA	Deoxy ribonucleic acid
DNP	Dinitrophenyl compound
HeNe	Helium-Neon
HSA	Human serum albumin
IR	Infra red
ITO	Indium tin oxide
LSPR	Localized surface plasmon resonances
MG	*N*-(2-mercaptopropionyl) glycine
MPA	Mercaptoproprionic acid
MPTES	γ-mercaptotriethoxysilane
OMCVD	Organo-metallic chemical vapor deposition
RIU	Refractive index unit
SAM	Self-assmbled monolayer
SEB	Staphylococcal enterotoxin
TE	Transverse electric (s-polarization, parallel to the glass slide)
THS	Thyroid stimulating hormone
TM	Transverse magnetic (p-polarization, perpendicular the glass slide)

Symbols

k-vector	Wave vector
n	Refractive index
w/v	Weight/volume
Δn	Refractive index change
λ	Wavelength

1 Introduction

Optical waveguides are long known for their properties to guide light from a laser source or a regular lamp along quite large distances [1–6]. The development of optical fibers and, in particular, their application in telecommunication have driven the development in integrated optics and applied integrated optics technology forward [7–9]. The key issue here is about the devices with very low losses: low losses to couple the light into the device, low losses during propagation, low losses during an active operation, and low losses when coupled out.

The introduction of a metal, which usually exhibits high absorptions, onto or close to a waveguide was a revolutionary idea. Critical voices only mentioned malfunctioning because of the extremely high losses expected and, therefore, the small possible propagation lengths.

Most optically based sensors work as immunosensors. Here, typically a surface is coated with a chemical component that has an affinity to the analyte molecule. This affinity can be as simple as a Coulomb attraction which would be very unspecific. The affinity can also be based on a highly sophisticated biological recognition reaction, such as the body-antibody recognition, which is highly specific and should avoid unspecific recognition and false positives. In all of these cases, the recognition leads to an increase in the film thickness on top of the sensor surface or an increase of the refractive index of this film. The interrogation of this binding event is, in most cases, done by using the change in the film thickness or refractive index on the surface.

2 Surface Plasmon Waveguide Coupling

2.1 Planar-Waveguide Structures

Regardless of the critical remarks about high optical losses, a couple of very successful attempts have been made in the late 1990s up to very recently with channel waveguides carrying a thin layer of gold for waveguide mode – surface plasmon coupling [10–22]. An overview article is available [23]. These devices were intended to excite hybrid modes of the surface plasmon with one or more waveguide modes mainly for sensor application (Fig. 1a). If the wave vectors (or k-vectors) of the waveguide mode and the surface plasmon can be matched precisely, a strong coupling between the two excitations can be achieved. The waveguide transmission decreases rapidly if the waveguide mode is strongly coupled to the plasmon. If one of the resonances is detuned, typically the surface plasmon resonance due to the binding of the analyte onto the gold functionalization, the wave vector match is changed or lost, therefore the coupling conditions are altered, and the waveguide transmission increases.

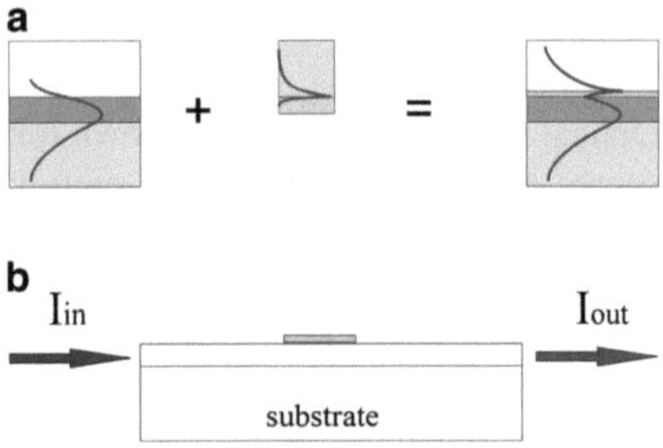

Fig. 1 (**a**) Schematic representation of waveguide mode–surface plasmon coupling via k-vector match into a hybrid mode. (**b**) Experimental scheme of a planar waveguide with a small gold spot on its surface for transmission experiments

Here, the idea was to combine the easy handling and characterization, in a transmission manner (Fig. 1b), without moving parts and bulky prisms, of a waveguide with the high sensitivity of the surface plasmon spectrum [24]. In addition, a gold layer offers easy access via sulfur chemistry and self-assembled monolayers [25, 26] for functionalization with receptors and recognitions sites. Furthermore, the combination of a multimode waveguide with a surface plasmon resonance offers a better characterization of the organic layers, both the receptors and the analyte, due to the multiple measurement possibilities. In a multimode experiment, both the refractive index and the film thickness can be measured [22].

2.2 *Surface Plasmon Waveguide Coupling in Optical Fibers*

Similar surface plasmon-waveguide devices have been proposed, designed, and constructed using optical fibers [27–37]. A brief review is also available [38]. The earliest contributions here were fiber optical transmission sensors, where in a small part of the fiber, the cladding was removed and replaced with a layer of gold or silver to allow for coupling of the waveguide mode and the surface plasmon. The coupling between the propagating waveguide mode and the surface plasmon on the metal dielectric interface on the outside of the fiber, also via wave vector match, was either achieved by propagating white light through the fiber and detecting a transmission spectrum at the end, or by launching a monochromatic wave from a laser source under a controlled coupling angle at the end-fire of the fiber (Fig. 2). In single wavelength cases, the transmission intensity versus coupling angle α (Fig. 2a) is recorded. The coupled surface plasmon is detected as an absorption feature in the wavelength transmission spectrum or as a decrease in the transmitted intensity in the angle spectrum. In all these approaches [27–33, 36, 37, 39], a sensor signal, a change in the position of the plasmon resonance in the respective spectrum, was picked up due to changes of the refractive index of the medium located around the sensor area, the metalized part of the fiber.

Recently, this technology was extended toward chemical sensing of target analyte molecules and dangerous pathogens like bacteria [34, 35]. Here, the

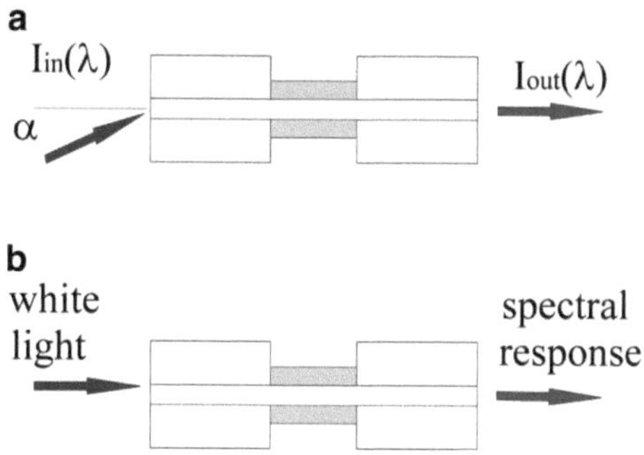

Fig. 2 Waveguide mode–surface plasmon coupling in fiber geometry with an area of removed cladding and a gold deposition therein. **a**) Experimental scheme for a single wavelength transmission experiment under controlled coupling angle α into the fiber. **b**) Experimental scheme for a spectral transmission experiment

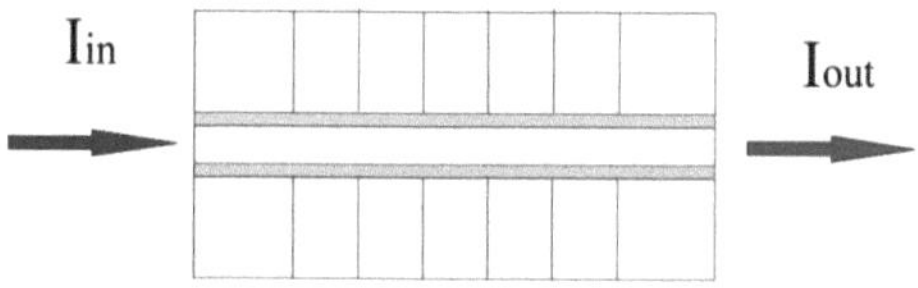

Fig. 3 Experimental scheme for a hollow optical fiber carrying a Bragg grating and a gold layer on the inner surface of the hollow fiber for waveguide mode–surface plasmon coupling and detection in transmission

metalized parts of the fibers were functionalized with a specific recognition site for the target to bind. Both the transmission spectra changes for a white light illumination and the transmission intensity changes with single wavelength excitation but with coupling angle dependence were investigated versus target concentration. Although multimode fibers were implemented in these fiber sensor developments, no use of the various modes was made.

A recent development was the surface-plasmon waveguide coupling with the help of a short-period Bragg grating in hollow optical fibers (Fig. 3). The metal layer was formed as the inner coating of the hollow fiber. Nemova and Kashyap [40] investigated theoretically the dependence between the effective refractive index of the surface plasmon resonance and the thickness of the metal layer and the refractive index of the medium to be sensed. They also investigated the influence on the grating period and length on the reflectivity behavior of the proposed Bragg reflector. They proposed a fiber transmission sensor interrogating the transmitted intensity spectrally upon a change in the refractive index on top of the metal coating.

3 Gold-Nanoparticle Devices

A parallel development in sensor technology is the use of gold nanoparticles (AuNPs) and their localized surface plasmon resonances (LSPR) [41] for sensing purposes. Mirkin et al. [42] showed how single nanoparticles (NPs) in solution can be clustered by a smart DNA recognition reaction triggered by a surface functionalization immobilized on them, which leads to an immense optical response in form of a color change or a change in the absorption spectrum.

These fundamental works and discoveries have stimulated an avalanche of investigations on metal nanoparticles in the gas phase, in solution, and on surfaces [43]. In this chapter, the focus will be on nanoparticles on surfaces and, in particular, on gold nanoparticles (AuNPs) on waveguides, both planar structures and fibers, as well as photonic crystals with the aim of sensor application.

3.1 *AuNPs on Planar Waveguides Without Application of LSPR*

One of the earliest, approaches to use AuNPs on a waveguide for sensor application was conducted by Busse at al. [44]. They grew AuNPs via organo-metallic chemical vapor deposition (OMCVD) [45–47] on top of the arms of a Mach-Zehnder interferometer (Fig. 4). The interferometer was used in its classical operation form outside the LSPR resonance. Extreme care was taken that the amount of AuNPs was small enough to avoid too high optical losses. Gold was used to circumvent the surface functionalization of the Mach-Zehnder interferometer arms with silanes, whereas thiols were applied to the AuNPs for functionalization. No features of the LSPR were exploited here. The device was based on the biotin–streptavidin recognition scheme [48] and was used to sense biotin-functionalized anti-THS, a thyroid stimulating hormone with a molecular weight of 75,000 Da. They could achieve a sensitivity of 2ng/cm^2 for the antibody. In a previous experiment with the identical Mach-Zehnder interferometer without AuNPs which was functionalized directly with silanes on its arms, the sensitivity was an order of magnitude larger: 0.2 ng/cm^2 [49]. The difference between these two interferometers was the amount of surface functionalization. In the directly functionalized case with the silanes, 100% of the waveguide surface was covered with the recognition sites, whereas in the AuNP case, only 10% of the waveguide surface could be functionalized, because otherwise the losses had been increased to a too high level to achieve proper interferometer operation.

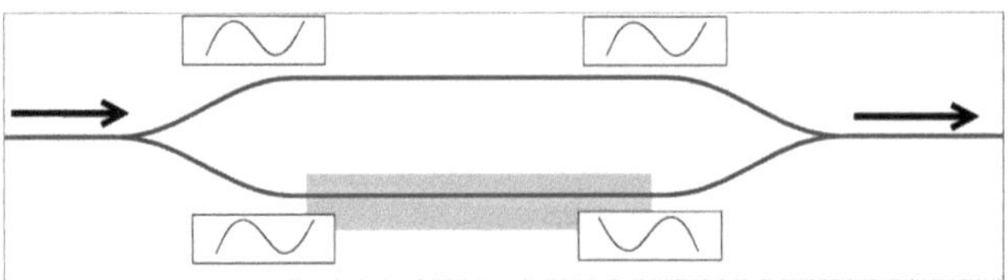

Fig. 4 Mach-Zehnder interferometer for classical interferometry. Here, one arm, the grey one, is functionalized with AuNPs to allow for a thiol surface functionalization for an immunoreaction, instead of a silanization procedure

3.2 *AuNPs on Planar Waveguides with Application of LSPR*

Simple transmission experiments to determine the spectral location of the LSPR with standard absorption spectrometers of random nanoparticles and ordered NP arrays suffer from a weak signal and a nonacceptable signal-to-noise ratio. This makes it very difficult to pick up the small spectral shifts of 10–40 nm in the LSPR spectral position, when the analyte binds to the recognition sites located on the nanoparticles [43]. Therefore, a combination of a spectrometer with a microscope is used to detect the LSPR in small NP arrays with a high signal-to-noise ration.

In order to overcome this signal-to-noise issues and to avoid these elaborate microscope-spectroscopy experiments, waveguides as substrates have been introduced. They allow integrating the LSPR absorption signal along an increased interaction length along the waveguide. The evanescent field of the waveguide serves here as the light source probing the LSPR spectral position.

In the sensor developments of AuNPs on waveguides, the experimental technology is based on transmission experiments and in some cases for optical fibers on a transmission-reflection experiment. Nearly all groups detect the LSPR spectral position by launching white light into a waveguide device and detect the spectral transmission dip, and its spectral shift due to a recognition event on the surface functionalization on the AuNPs (Fig. 5).

In 2004, Storhoff et al. [50] used the distance-dependent optical LSPR behavior found in solution upon particle clustering in combination with a quasi optical waveguide: a glass slide which shows substrate modes or periodic spots of total internal reflection along the propagation direction. No polarization dependence was studied. The authors prepared two sets of DNA-functionalized 40–50 nm (probably spherical) AuNPs in solution exhibiting a LSPR band at 528 nm. The two DNA strands in combination were complementary to a target DNA, so that a hybridization reaction could occur upon mixing the target DNA with the two AuNP samples in solution. After the hybridization reaction, the LSPR band occurred around 560 nm, indicating, thereby, the clustering of the AuNPs initiated by the target DNA. The sensing was carried out by spotting both AuNP samples into a spot and spotting the mixed sample (both AuNPs and the target DNA) into another spot onto the pseudo waveguide. The hybridization reaction was detected visually by the naked eye from the color induced due to the evanescent

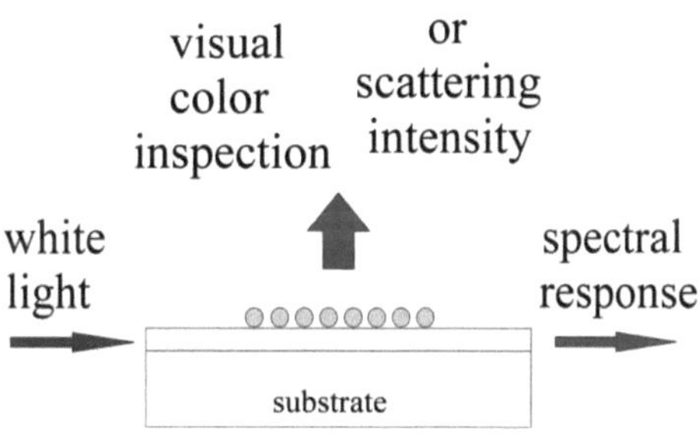

Fig. 5 AuNPs immobilized on planar-waveguide structures not only for transmission absorption spectroscopy, but also for visual inspection and measurement of scattered intensity at a 0° geometry from the waveguide surface

illumination of the AuNPs in the spots. The spots of individual NPs appeared green, whereas the clustered AuNPs appeared orange. In order to increase the sensitivity, dextran sulfate was used to enhance the hybridization kinetics. The colors of the spots were determined reproducibly with a color-sensitive CCD camera by a quantified analysis of the red channels. In a final experiment, Storhoff et al. measured the spectrum of scattered photons of the AuNPs with a diode array detector. Here, they could achieve an increase in scattered intensity and a spectral shift with the increase of the target DNA.

Broadband, time-resolved, and polarization-dependent spectroscopy with a 50-μm thick quasi-waveguide exhibiting multiple reflections was performed by Qi et al. [51]. A silica fiber was used to couple white light into the glass plate. At the end, the light was analyzed by a multichannel CCD detector (400–800 nm). The AuNPs (probably spherical), synthesized in aqueous solution by reducing $HAuCl_4$ with glucose and fructose, showed a LSPR in aqueous solution at $\lambda = 533$ nm. These AuNPs were immobilized on the glass slide with the aid of a hemoglobin layer on the glass slide and attractive Coulomb interaction between the AuNPs and the hemoglobin. The authors monitored the self-assembling of the solution-born AuNPs onto the glass slide in TE (s-polarization, parallel to the glass slide) and in TM (p-polarization, perpendicular the glass slide) direction. For the TE mode, they find an increasing LSPR absorption band at 548 nm, whereas for TM the peak is located at 539 nm and increases accordingly with time, both slightly different from the solution spectrum due to the new anisotropic environment. From these spectral row data, they were able to conclude a diffusion-driven self-assembly process for the AuNPs. Surprisingly, the LSPRs shifted and broadened the TE mode to slightly higher wavelength by exchanging the gold colloid solution to pure water. In the TM case, no shift was observed but only a slight line broadening. The reason for this behavior might be due to a change in the charge distribution or a change in the conformation of the hemoglobin.

Hamamoto et al. [52] also used a quasi waveguide (without utilizing polarization dependences), a glass slide with multiple total internal reflections, but they coupled the light into the glass slide under a controlled angle at the tilted end-face of the slide (similar to Fig. 2a). The transmitted light is analyzed via a CCD camera, whereas the scattered light is observed in the middle of the device under a 90° geometry with photodiode. AuNPs of various sizes and forms were grown onto the surface of the glass slide. The analyte solution, a glucose solution with a concentration of up to 16% w/v, is placed on top of the immobilized AuNPs. The authors measured the extinction spectra (not explained how) of their system as a function of the glucose concentration and found the typical red shift of the LSPR resonance with increasing glucose concentration due to an increase in the refractive index of the environment. The scattered intensity (wavelength unknown), detected with the photodiode under 90° geometry, showed a stepwise intensity increase with a stepwise glucose concentration enhancement. The relationship between the scattered intensity and the glucose concentration was linear. The transmitted intensity at the end of the glass plate taken at $\lambda = 530$ nm, which is located close to the LSPR band maximum of the 16% w/v concentration with the CCD camera, was determined as a function of the coupling angle into the glass slide. In this angle spectrum, intensity maximums occurred being dependent on the glucose concentration: higher the glucose concentration the lower the "resonance" angle. The physical reason for this resonance remains unexplained. Also in this case, the author found a strictly linear behavior between the angle position of the "resonance" and the glucose concentration.

AuNPs with pancake-like shape (aspect ratio = width : height = 3 : 2, with sizes in the order of a few nanometer) were grown by OMCVD [44–47] onto a monomode ion exchanged channel waveguide by Aliganga et al. [53]. Due to the statistical process of OMCVD growth, both the size and the distance of the AuNPs to their nearest neighbors are statistical and random. This leads to a relatively broad LSPR. The authors performed a spectroscopic examination of the AuNPs in air with the TE and the TM mode by launching white light into the polished end-face of the channel waveguide and feeding the transmitted light into a spectrometer. Because of the shape dissymmetry of the immobilized AuNPs, they found a broad LSPR for the TE mode at $\lambda = 499$ nm, whereas the broad TM mode showed the resonance $\lambda = 493$ nm. The broad peaks in both polarization directions depict the statistical size distribution in both directions: widths and height. The immobilization of a 1-dodecalthiol self-assembled monolayer led the TM mode (measured in air and probing the direction perpendicular to the waveguide surface) to a LSPR at $\lambda = 503$ nm, a shift of 10 nm.

However, for the TE investigations (in air and by probing parallel to the waveguide surface), a different behavior was found. The peak shifted by 11 nm from $\lambda = 499$ nm to $\lambda = 510$ nm, but in addition, a strong shoulder evolved in the near IR between 700 and 750 nm. This was ascribed to the distance-dependent optical LSPR behavior of a part of the AuNPs as found in solution upon particle clustering. Here, the clustering is triggered by the SAM and called "optical cross-talk." The cross-talk is induced when the evanescent fields of the AuNPs start to overlap, generated by the increased refractive index of the SAM and by pulling out the evanescent fields slightly.

This interpretation was confirmed by theoretical investigations and confirming measurement by Rooney et al. [54], who approximated these pancake-shaped AuNPs by half spheres. Here AuNPs were grown by OMCVD on ion-exchanged monomode slab waveguides. The polarization was systematically tuned in 15° steps from pure to TE to pure TM in air. In the case of the uncoated AuNPs, only a change in the background could be detected, whereas in the case of a 1,8 octanedithiol SAM coating, a second peak was found at 755 nm confirming the data of Aliganga et al. The DDA simulations were carried out with pairs of half-spheres illuminated with s- and p-polarized light.

Indeed, the simulations show a large shift of the LSPR upon optical cross-talk introduced by a thin layer of dielectric material onto the AuNPs. This cross-talk is distance dependent. The DDA simulations were also carried out with water as the AuNP environment. The simulations show that in the case of water, the peak shift is less pronounced. A comparison in the cross-talk behavior of rows (2–4 AuNPs) of these half-sphere particles with rows of spherical AuNPs located on a waveguide

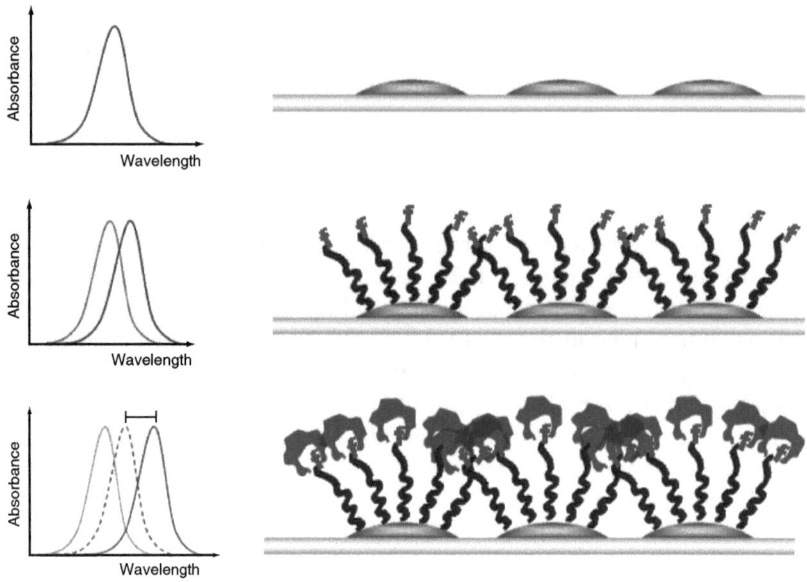

Fig. 6 Proposed AuNP cross-talk sensor. AuNPs with a distinct distance are immobilized on a waveguide. In s-polarization, they show an LSPR. After functionalization with a recognition moiety for a particular analyte molecule, a tiny red shift is expected for the LSPR. When the analyte binds the optical cross-talk sets in and the LSPR shifts strongly to higher wavelengths

surface have been carried out theoretically by Hashemi Rafsanjani [55]. Here, it turns out that this optically induced clustering or SAM-induced cross-talk behavior can only be achieved by semi-spherical AuNPs. The evanescent field between the half-spheres is enhanced by a factor of 10^5 due to the sharp edge directly on the waveguide surface, leading to such a pronounced appearance of the cross-talk peak. In the case of complete spheres, no cross-talk peaks could be observed theoretically at comparable distances.

The authors, therefore, proposed a very sensitive sensor based on this cross-talk peak shift (Fig. 6). AuNPs will be immobilized on a waveguide, which show an average spacing large enough to avoid the appearance of a cross-talk peak when a surface functionalization with a specific recognition site for a target molecule is immobilized. When the target is binding, the optical clustering is switched on and the cross-talk peak appears. The "intensity" of the cross-talk peak should be directly related to the concentration of the target.

AuNPs of 5 nm and 20 nm (probably spherical) were synthesized and immobilized with the aid of γ-mercaptotriethoxysilane (MPTES) or 3 aminopropyltriethoxysilane (APTES) on a quartz plate serving as a quasi waveguide with multiple reflections for absorption spectroscopy by Nakanishi et al. [56]. The authors found a LSPR with standard absorption spectroscopy of the 4 nm AuNPs in toluene at $\lambda = 530$ nm. The same LSPR band position was found for the 20 nm AuNPs in water modified with a layer of mercaptoproprionic acid (MPA). The 4 nm AuNPs

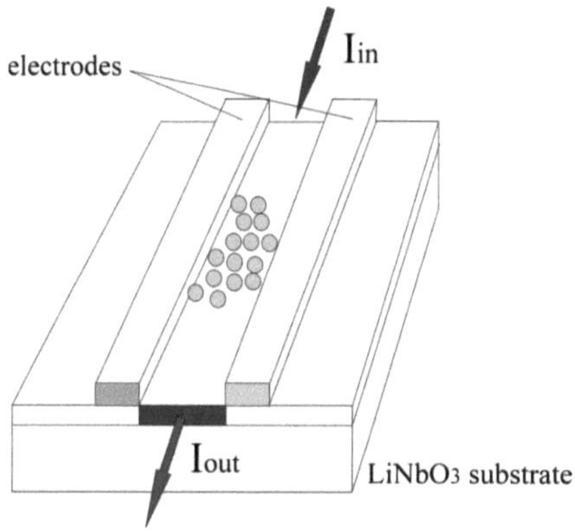

Fig. 7 Experimental scheme of a $LiNbO_3$ waveguide with an electro-optical interrogation. Here, two electrodes are present, which allow to apply a potential across the electro-optically active waveguide. This leads to a change in the k-vector of the propagating waveguide mode. The transmission of either a single wavelength or a white light spectrum can be measured, as well as the phase between the transmitted single wavelength and a reference beam can be measured

immobilized via MPTES showed a broad LSPR spectrum with a maximum at 580 nm. The 20 nm MPA-modified AuNPs immobilized via APTES showed a severe broadening with a very strong "shoulder" in the near IR from 700 to 900 nm and peak at 580 nm as well. The shoulder was explained by particle coupling. They proposed to use this LSPR shifts and spectral broadening for biotin–streptavidin recognition.

A combination of a nonlinear optical approach, an electro-optical modulation (Fig. 7) with a Ti-diffused $LiNbO_3$ channel waveguide was presented by Wang and Lin [57, 58]. $LiNbO_3$ is known to be an excellent material for an electro-optical response. The AuNPs (probably spherical) with diameters of 2 or 22 nm were immobilized via 3-aminopropyltrimethoxysilane on a spot on the channel waveguide. On both sides beside the AuNP spot electrodes were placed. The electrodes were covered with SiO_2 to avoid direct electrical contact to the AuNPs. The electrical field was strong in the $LiNbO_3$ waveguide but week in the AuNP area. The wave vector of the propagating light beam in the waveguide was modulated by applying a voltage on the electrodes. The LSPR of the AuNPs was excited when the wave vector of the propagating photons matches the wave vector of the immobilized AuNPs. The AuNP wave vector depends on the direct environment around the AuNP. The AuNPs were coated via self-assembling with human serum albumin (HSA). The analyte was a β-blocker that can bind to HSA to form a protein–drug complex. The authors launched a HeNe laser beam at $\lambda = 632.8$ nm into the channel waveguide as a TM mode and measured the transmitted intensity with a power meter and gave the results as a loss in dB. Alternatively, they launched white light into the waveguide and analyzed the transmitted light with a spectrometer. Both experiments were conducted under different potential conditions across the waveguide. The spectral analysis showed the AuNP LSPR band between 617 and 620 nm when no voltage is applied depending on the width of the waveguide (unknown AuNP size). A 20-μm-wide waveguide showed the higher LSPR wavelength, whereas a 60-μm-wide waveguide depicted the lowest LSPR resonance wavelength. The LSPR band shifted linearly to the red with increasing voltage; revision of the polarity lead to a linear blue-shift of the LSPR for all chosen waveguide widths.

In the single-wavelength transmission experiment, at 632.8 nm, the authors found a change in the transmitted intensity at a fixed voltage with changing beta-blocker concentration. They found a linear dependency on the relative transmitted intensity in dB (ratio of the transmitted intensity with β-blocker normalized to the transmitted intensity without any β-blocker but with the pure HAS functionalization) versus applied voltage. The experiments were carried out with various β-blocker concentrations. The slopes were all identical, but the overall loss increased nearly linearly with increasing β-blocker concentration.

In a second serious of experiments [58], the authors interrogated the phase difference between the light beam from the HeNe laser, which was transmitted through the device under potential, and a direct beam from the HeNe laser. Technically, they achieved this by splitting the HeNe laser beam into two polarizations with individual acousto-optical modulation and used phase-sensitive lock-in technology to determine the phase relation between both. At a given potential across the waveguide, they found an increase in the phase difference with increasing β-blocker concentration. For the various waveguide widths they found a linear increase of the phase difference with the increase in β-blocker concentration. The wider the waveguide was, the larger the overall phase shift was. The phase shift also showed a strictly linear increase with increasing voltage across the waveguide. Here, the slope varied with the β-blocker concentration.

3.3 AuNPs and Photonic Crystal Waveguides

A submicrometer-sized silicon-integrated structure is used to detect the presence of a few down to a single 10-nm-sized AuNPs by Schmidt et al. [59]. A strong light confinement in a nano-cavity is achieved to enhance the extinction cross-section of the AuNPs at a wavelength regime of 1450–1600 nm. The structure used is a 1D photonic crystal made out of a high-index silicon channel waveguide with 200 nm diameter holes filled with silicon dioxide, exhibiting a lower refractive index (Fig. 8). This pattern creates a distributed Bragg reflector. In the middle of the waveguide structure, a 100 nm diameter hole, which is also filled with SiO2 is located. This defect creates a local discontinuity in the propagating TE field, increasing the field strength in the middle of the device – the nano-cavity – by a factor of 350 in comparison to the "regular" waveguide. Because of the presence of the Bragg gratings, this device acts as a microcavity. The authors launch white light into the waveguide micro cavity and measure the transmission spectra of the device as a function of the diameter of the central nanocavity. They find transmission spectra with a relative sharp peak which shifts to higher infrared wavelength linearly with increasing nano-cavity diameter. When placing AuNPs into the nanocavity area, the transmission peak shifts to higher wavelength. They can detect a small peak shift even with only a single NP within the nanocavity. The authors also performed a transmission experiment (at unknown wavelength conditions) where they systematically increased the NP density on the nanocavity from 0 to 0.26 particles/μm^2. The found the transmission to decrease with increasing NP concentration. The authors propose to use this technology for sensing biologically relevant molecules with single AuNPs.

A much simpler approach was taken by, Hojjati et al., Linden et al., Gantzounis et al., Jiang et al. [60–63]. They have placed AuNPs in regular pattern on slab waveguides modeled or fabricated out of ITO or glass (Fig. 9). The device characterization was either carried out as a transmission experiment with a light beam incident perpendicular on the AuNP / waveguide structure and studying the principle physical effects [61, 62] or as a waveguide transmission experiment with a view toward sensor application [60, 63]. An overview article on the nonsensing activities is available [64].

Briefly, square AuNP pattern were fabricated on 140 nm thick ITO waveguides with a lateral spacing of 300 nm in one direction and a systematically changed

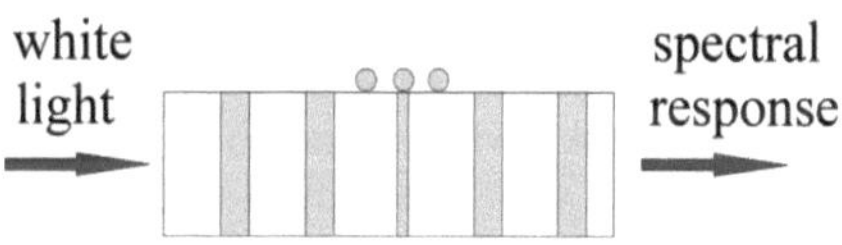

Fig. 8 Experimental scheme of a spectral transmission experiment of a photonic crystal Bragg reflector micro-cavity fabricated out of refilled holes in silicon. The central refilled hole acts on a nanocavity. AuNPs are immobilized in and around the central nanocavity

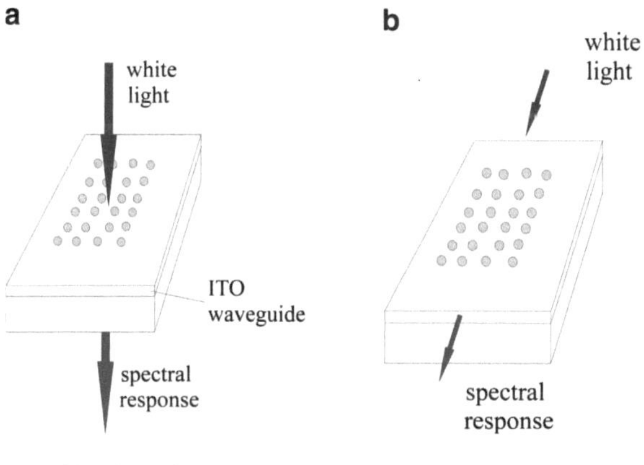

Fig. 9 Planar-waveguide photonic crystals by a periodic pattern of AuNPs on the waveguide surface. (**a**) Experimental scheme for a spectral transmission experiment under 0° geometry through the AuNP pattern and the waveguide structure. (**b**) Experimental scheme of a spectral transmission experiment applying waveguide modes

spacing between 350 and 475 nm in the perpendicular direction. The AuNPs were roughly 100 nm in diameter and 20 nm in thickness. The white light was incident under 0° onto the waveguide structure and the transmitted light was analyzed spectroscopically (Fig. 9a). The authors found "dips" in the LSPR band. The spectral position of these dips changed systematically to smaller photon energy with increasing pattern spacing. These dips were interpreted as a coupling of part of the white light via k-vector matching into TE modes of the waveguide. By performing the same experiments with a waveguide thinner than the cut-off thickness (ITO of 30 nm thickness), the dips did not appear. The dips could be reinitiated by adding a refractive index liquid with a high refractive index brining the waveguide above the cut-off. No application for biosensing is proposed so far taking this "spectral hole burning" into account.

Jiang et al. simulated the transmission of white light propagating as a waveguide mode through a 140-nm-thick ITO waveguide carrying an AuNP pattern with a systematically varying spacing between 300 nm and 500 nm in transmission direction and perpendicular (Fig. 9b). The AuNPs were 100 nm high and had a diameter of 100 nm. First, they studied the absorption spectrum for a single AuNP located on the ITO waveguide for the TE and the TM mode. They found the LSPR as a sharp peak at 690 nm for the TE mode which only probed the nanoparticle along its diameter direction (parallel to the waveguide). Because TM modes probe mainly the out of plain direction of the particle on the waveguide but also have a small field component in propagation direction probing again the diameter of the AuNP, two LSPR absorption peaks were found (680 nm and 600 nm). In order to investigate the photonic crystal effect the absorption spectra for systematically varying pattern were simulated for TE modes. The LSPR shifted toward the blue from 690 nm to 665 nm with increasing pattern spacing in propagation direction. When the spacing in both directions was varied simultaneously the spectra changed quite dramatically: from two LSPR to one LSPR with shifting maximum position. The reason for this complicated behavior lies in the in-phase addition of the grating vector through the emerging AuNP grating.

The influence of the waveguide thickness (from 125 nm to 225 nm) on the LSPR was investigated with a fixed pattern structure. Here clearly a spectral red shift of the LSPR with increasing waveguide thickness was found. Finally they simulated the effect of a dielectric coating at a fixed AuNP pattern simulating a sensor response. The refractive index of the coating was chosen as $n = 1.5$, typically for an organic material, and the thickness was varied from 5 to 40 nm. They found a shift of the LSPR with increasing coating thickness of up to 13 nm.

A first experiment has been performed with a AuNP pattern located on a mono mode ion exchanged glass waveguide in air [60]. The AuNPs were aligned in a square pattern with a center to center spacing of 500 nm. The AuNPs were 100 nm in diameter and 75 nm in height. Waveguide transmission experiments with the TE_0 and the TM_0 mode were carried out and simulated. Spectra with a variety of regular absorption features were found both in the experiment and in the simulations for both polarization directions. The strongest feature around 625–640 nm (depending on polarization) is interpreted as the LSPR whereas the other absorption features are interpreted as incomplete band gaps. The authors propose a sensor scheme based on a coupling between the LSPR and a strong band gap to produce a degenerated state which might lead to a band splitting. The experimental conditions for such a band splitting seem to be quite tide so that a small detuning, e.g., due to the binding of minute amounts of analyte would destroy that degenerated state.

3.4 AuNPs on Optical Fibers with Application of LSPR

In 2000, a light beam from a white light source filtered by a monochromator was launched into an optical fiber for transmission absorption spectroscopy by Meriaudeau et al. [65]. The optical fiber carried AuNPs (shape unknown) with a size of 4 nm at its exit end-face (Fig. 10). The end-face of the optical fiber was immersed in a cuvette that could be rinsed with liquids of different refractive index. The LSPR of the AuNPs in air ($n = 1$) was found at 530 nm and shifted systematically and linearly to higher wavelength with the increase of the refractive index of the AuNP surrounding liquid. They achieved a LSPR shift of about 100 nm for a Δn of 0.7. The authors described this as a first fiber sensor for refractive indices of solutions.

A reflection-based experiment with AuNPs (probably spherical) with a size of 20 nm located at the end-face of an optical fiber was carried out by Mitsui et al. [66]. They recorded the absorption spectra in a reflection mode using a fiber coupler to guide the reflected spectrum to a spectrometer. In an alternative setup, they launched light at 850 nm, which is spectrally located close to the maximum of the LSPR of the used AuNPs, from a light-emitting diode into the fiber. The intensity of the back reflected light is analyzed via lock-in technology. The nanoparticles were functionalized with biotin and a recognition experiment with avidin was performed. The authors find a very small shift in the LSPR among the uncoated AuNPs in water (at 550 nm), the biotin-coated AuNPs, and the biotin–adenin-coated AuNPs. However, the absorption increased due to avidin binding.

The reflected intensity of the 850 nm photons were investigated with uncoated AuNPs and with a systematic change of the refractive index of the medium around the AuNPs. A strongly linear increase of the reflected intensity with increasing refractive index was found with a slope of 29.9 per refractive index unit (RIU). The intensity change is due to the LSPR shift and a fixed probing wavelength. With increasing refractive index, the LSPR moves away from the probing wavelengths and therefore allows more photons to be reflected and not absorbed. Finally, they

Fig. 10 Immobilized AuNPs on the end fire of an optical fiber for transmission and reflection measurements

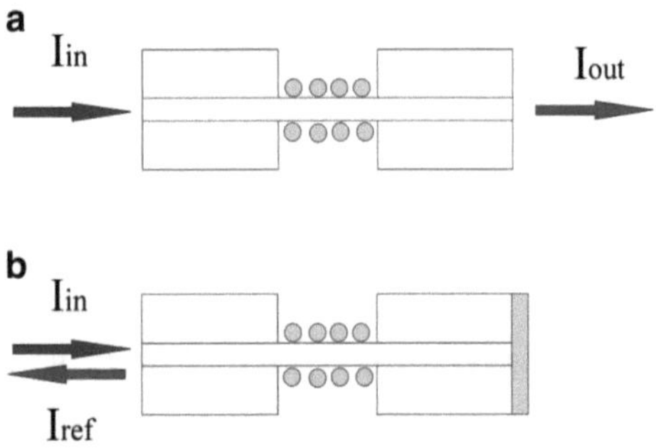

Fig. 11 On an optical fiber, the cladding is removed along a particular length and AuNPs are immobilized there, right on the waveguide surface. (**a**) Experimental scheme for a transmission experiment. (**b**) Experimental scheme for a reflection experiment with a silver mirror at the end face

conducted a kinetic avidin-binding experiment where they followed the change of reflected intensity, an increase, with time after injecting the avidin.

A theoretical study was performed by Sharma and Gupta [67] on fiber optical devices in transmission. The cladding of the fiber was removed on a part of the fiber and AuNPs with a shell of a nonabsorbing, dielectric material were deposited there, directly onto the waveguide (Fig. 11a). The authors studied the spectral behavior of the transmitted power, the sensitivity as well as the signal-to-noise ratio with respect to the AuNP size, the thickness of the shell, and its refractive index. They find the following dependencies. (a) The signal-to-noise ratio increases significantly with increasing AuNP size, but the sensitivity does not show an important dependency. (b) The amount of AuNPs is an important factor: the signal-to-noise ratio and the sensitivity increase with the amount of AuNPs. (c) The dielectric constant of the shell material has an influence on the signal-to-noise ratio and the sensitivity, however, in reverse direction. The signal-to-noise ratio decreases and the sensitivity increases with increasing dielectric constant. Similar results were found by these authors theoretically by using silver–gold alloy NPs [68].

Chau et al. [69] demonstrated both reflection and transmission fiber optical sensing (Fig. 11) with 7-nm-sized AuNPs immobilized directly onto the active waveguide in an area of removed cladding and being immersed in a cuvette. The length of the removed cladding and therefore the amount of immobilized AuNPs were systematically varied. In the case of the reflection experiments, the end fire of the fiber was polished and coated with silver to produce a high-quality mirror. For the sensor characterization, the authors launched a 635 nm diode laser beam into the fiber and measured the relative reflected or transmitted intensity. The relative intensity was determined by normalizing the measured intensity to a standard, e.g., the intensity achieved for the blank fiber. The authors first tested the transmitted intensity versus the refractive index of a medium surrounding the AuNPs with respect of a removed cladding length between 0.5 cm and 5 cm. They found a nonlinear decrease in relative intensity up to 70% with increasing refractive index (for a $\Delta n = 0.09$) but an unsystematic dependency on the removed cladding length. The authors attribute this behavior to a change in the numerical aperture of the fiber with changing refractive index in the cladding area. In the comparable reflection experiment, a nearly identical behavior was found. For further experiments, they chose the 2 cm removed cladding in reflection mode interrogation.

After the AuNPs were functionalized (on a glass slide for spectral analysis) with *N*-(2-mercaptopropionyl) glycine (MG) which is sensitive to the binding of divalent metal ions, such as Ni^{2+} or Pb^{2+}, absorption spectra were taken, showing only a very minor shift of the LSPR between AuNPs with and AuNPs without the presence of Pb^{2+} ions, but a slight increase in the LSPR peak height. The MG-functionalized AuNPs on the fiber in reflection mode yielded a 2% relative intensity change in a very linear fashion with increasing Ni^{2+} concentration up to 1.0 mM. The same reflection and absorption spectroscopy experiments were carried out with biotin-functionalized AuNPs. Here, the LSPR showed an absorption peak increase without any shift upon stretavidin binding, whereas the reflection intensity measurement showed a linear decrease up to 4% for a 10 nM streptavidin solution. To prove the capability of the device as an immunosensor, the AuNPs were functionalized with an antibody against staphylococcal enterotoxin (anti-SEB). The binding of the SEB yielded a linear relative intensity change of about 2.5 % for a 100 ng/ml SEB concentration.

The same group presents a slightly more sophisticated fiber sensor operating on 8-nm-sized AuNP located in the area of removed cladding on an optical fiber carrying a long-period grating [70]. The grating period was chosen to be about 550 nm leading to a "forbidden" transmission band between 1510 nm and 1600 nm. The spectral response of the AuNPs in combination with the Bragg grating was detected in a fiber transmission experiment by launching "white" light from a light source emitting near infrared wavelengths into the fiber and analyzing the transmitted light by a spectrometer. The authors first tested the influence of the refractive index surrounding of the AuNPs and found the peak wavelength of the forbidden transmission band shift to lower wavelength with increasing refractive index. The spectral band shifted from 1521.7 nm at $n = 1.33$ (water) to 1520.55 for $n = 1.385$ (sucrose solution). In addition, the dip depth decreased with increasing refractive index. The AuNPs were functionalized with a dinitrophenyl compound (DNP) which is able to bind an anti-DNP. The spectral position of the transmission dip did not change upon rinsing anti-DNP into the cuvette with increasing concentrations. However, the dip depth changed and, therefore, the binding reaction could be followed with changing concentration.

3.5 *AuNPs and Photonic Crystal Optical Fibers*

Yan et al. [71] have used a hollow photonic crystal fiber bundle to immobilize multifaceted AuNPs of 100–200 nm diameters within the hollow structure (Fig. 12). They launched the excitation light at 633 nm for Raman spectroscopy onto the end fire of the photonic crystal fiber with a $\times 50$ objective. The same objective lens was used to collect the scattered light for Raman spectroscopy. The sensing material or analyte was rhodamine B immobilized and dried within the photonic fiber onto the AuNPs. The Raman spectrum showed a clear surface enhancement in the presence of the AuNPs. Nevertheless the authors used a photonic crystal fiber, no band gap

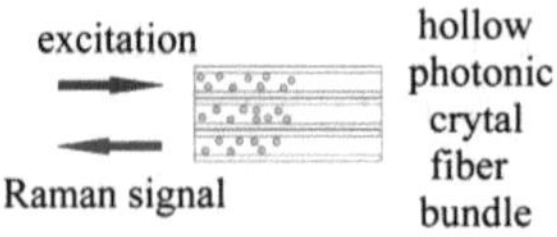

Fig. 12 Experimental scheme of a surface-enhanced Raman experiment with a hollow photonic fiber bundle. The AuNPs are immobilized on the inner surfaces of the hollow fibers and are illuminated by a focused excitation laser beam. The Raman signal is collected on the same side

was present at 633 nm, leading to weak light confinement in the fiber. The authors conclude that the use of a hollow photonic crystal fiber is a robust sensor platform which involves many particles for surface-enhanced Raman sensing.

4 Concluding Remarks

In the last 10 years, biosensors were proposed and realized both on planar waveguides and on optical fibers, classical fibers and hollow fibers, via surface plasmon–waveguide coupling and via AuNP LSPR interaction on waveguides or quasi-waveguides. The AuNP LSPR was also used on photonic crystal devices again in fiber geometry and in slab or channel waveguide form. The detection technology reaches from simple one-wavelength transmission measurement or reflection methods to spectral investigations and the implementation of nonlinear optical effects, such as electro-optical modulation technology. Heterodyne phase interrogation was used as well. Surface plasmon–waveguide coupling and LSPR of AuNPs was used to change the optical characteristics of Bragg gratings.

Nevertheless, many optical possibilities and possible combinations were tested; the advantages multimode waveguides and fibers offer have not yet been fully implemented. Polarization studies were reported, but are not yet fully appreciated for sensitivity enhancement and specificity issues.

The devices were applied not only to simply determine the refractive index of a liquid but also to detect a particular analyte molecule by an immunosensing reaction. Sensitivities down to zeptomole [50] quantities were reported.

In the future, we will see developments involving surface enhanced Raman scattering technologies in combination with AuNPs and waveguides as well as combinations of immobilized and solution-born AuNPs and functional bridges in-between them. It can also be expected that applied medical research, namely the detection of antigens, enzymes, and proteins in body fluids, will benefit from the sensor developments with its extreme sensitivities.

References

1. Adams MJ (1981) An introduction to optical waveguides. Wiley, New York
2. Hunsperger RG (1995) Integrated optics: theory and technology, 4th edn. Springer, Berlin
3. Marcuse D (1991) Theory of dielectric optical waveguides, 2nd edn. Academic, Boston

4. Snyder AW, Love JD (1983) Optical waveguide theory. Chapman and Hall, London
5. Syms R, Cozens J (1992) Optical guided waves and devices. McGraw-Hill, London
6. Tamir T (1979) Integrated optics, from topics in applied physics, 2nd edn. Springer, Heidelberg
7. Jones WB (1988) Introduction to optical fiber communication systems. Holt, Rinehart, and Winston, New York
8. Okoshi T, Kikuchi K (1988) Coherent optical fiber communication. Kluwer Academic, Dordrecht
9. Säckinger E (2005) Broadband circuits for optical fiber communication. Wiley-Interscience, Hoboken
10. Brecht A, Gauglitz G (1997) Lable free optical immunoprobes for pesticide detection. Anal Chimica Acta 347:219–233
11. Ctyroky J, Homola J, Skalsky M (1997) Modeling of surface plasmon resonance waveguide sensors by complex mode expansion and propagation method. Opt Quantum Electron 29:301–311
12. Ctyroky J, Homola J, Skalsky M (1997) Tuning of spectral; operation range of waveguide surface plasmon resonance sensor. Electron Lett 33:1246–1248
13. Dostalek J, Ctyroky J, Homola J et al (2001) Surface plasmon resonance biosensor based on integrated optical waveguide. Sensors Actuators B Chem 67:8–12
14. Harris RD, Wilkinson JS (1995) Waveguide surface plasmon resonance sensors. Sensors Actuators B Chem 29:261–267
15. Homola J, Ctyroky J, Skalky M et al (1997) A surface plasmon resonance based integrated optical sensor. Sensors Actuators B Chem 38–39:286–290
16. Lavers CR, Wilkinson JS (1994) A waveguide-coupled surface-plasmon sensor for an aqueous environment. Sensors Actuators B Chem 22:75–81
17. Matsushita T, Nishikawa T, Yamashita H et al (2008) Development of a new single-mode waveguide surface plasmon resonance sensor using a polymer imprint process for high throughput fabrication and improved design flexibility. Sensors Actuators B Chem 129: 881–887
18. Mouvet C, Harris RD, Maciag C et al (1997) Determination of simazine in water samples by waveguide surface plasmon resonance. Anal Chimica Acta 338:109–117
19. Weiss MN, Srivastava R, Groger H et al (1996) A theoretical investigation of environmental monitoring using surface plasmon resonance waveguide sensors. Sensors Actuators A Phys 51:211–217
20. Weisser M, Menges B, Mittler-Neher S (1998) Multimode integrated optical sensor based on absorption due to resonantly coupled surface plasmons. Proc SPIE 3414:250–256
21. Weisser M, Menges B, Mittler-Neher S (1999) Refractive index and thickness determination of monolayers by multi mode waveguide coupled surface plasmons. Sensors Actuators B Chem 56:189–197
22. Weisser M, Käshammer J, Matsumoto J et al (2000) Adenin - uridin base pairing at the water-solid-interface. J Amer Chem Soc 122:87–95
23. Mittler S (2006) Gold coated waveguide sensors. In: Grimes CA (ed) Encyclopedia of sensors, 4th edn. American Scientific, California, pp 239–259
24. Abdulhalim I, Zourob M, Lakhtakia A (2008) Surface plasmon resonance for biosensing: a mini review. Electromagnetics 28:214–242
25. Schreiber F (2000) Structure and growth of self-assembled monolayers. Prog Surf Sci 65:151–256
26. Ulman A (1991) An introduction to ultrathin films. Academic, San Diego
27. Cherif K, Hleli S, Abdelghani A et al (2002) Chemical detection in liquid media with a refractrometric sensor based on a multimode optical fiber. Sensors 2:195–2004
28. Gupta BD, Sharma AK (2005) Sensitivity evaluation of a multilayered surface plasmon resonance-based fiber optical sensor: a theoretical study. Sensors Actuators B Chem 107:40–46

29. Ika M, Seki A, Watanabe K (2004) Hetero-core structured fiber optic surface plasmon resonance sensor with silver film. Sensors Actuators B Chem 101:368–372
30. Januts NA, Baghdasaryan KS, Nerkarayan KV et al (2005) Excitation and superfocusing of surface plasmon polaritons on silver coated optical fiber tip. Opt Commun 259:118–124
31. Jorgenson RC, Yee SS (1993) A fiber optical chemical sensor based on surface plasmon resonance. Sensors Actuators B Chem 12:213–320
32. Khosravi H, Tilley DR, Loudon R (1991) Surface polariton in cylindrical optical fiber. J Opt Soc Am A 8:112–122
33. Lin BW, Jaffrezic-Renault N, Gagnaire A et al (2000) The effect of polarization of the incident light-modeling and analysis of a SPR multimode optical fiber sensor. Sensors Actuators A Phys 84:198–2004
34. Lin HY, Tsao YC, Tsai WH et al (2007) Development and application of side-polished fiber immunosensor based on surface plasmon resonance for the detection of Legionella pneumophila with halogens light and 850 nm-LED. Sensors Actuators A Phys 138:299–305
35. Lin HY, Tsai HW, Tsao YC et al (2007) Side-polished multimode fiber biosensor based on surface plasmon resonance with halogen light. Appl Opt 46:800–806
36. Nemova G, Kashyap R (2006) Fiber Bragg grating assisted surface plasmon-polariton sensor. Opt Lett 31:2118–2120
37. Ronot-Trioli C, Trouillet A, Veillas C et al (1996) Fiber optic chemical sensor based on surface plasmon monochromatic excitation. Anal Chemica Acta 319:121–127
38. Homola J, Lee SS, Gauglitz G (1999) Surface plasmon resonance sensors: review. Sensors Actuators B Chem 54:3–15
39. Ronot-Trioli C, Trouillet A, Veillas C et al (1996) Monochromatic excitation of surface plasmon resonance in an optical-fiber refractive index sensor. Sensors Actuators A Phys 54:589–593
40. Nemova G, Kashyap R (2006) Modeling of plasmon-polariton refractive index hollow core fiber sensors assisted by a fiber Bragg grating. J Lightwave Technol 24:3789–3796
41. Kreibig U, Vollmer M (1995) Optical properties of metal clusters. Springer, Berlin
42. Mirkin CA, Letsinger RL, Mucic RC et al (1996) A DNA-based method for the rationally assembling nanoparticles into macroscopic materials. Nature 382:607–609
43. Willets KA, van Duyne RP (2007) Localized surface plasmon resonance spectroscopy and sensing. Annu Rev Phys Chem 58:267–297
44. Busse S, Käshammer J, Krämer S et al (1999) Gold and thiol surface functionalized integrated optical Mach-Zehnder interferometer for sensing purposes. Sensors Actuators B Chem 60:148–154
45. Käshammer J, Wohlfart P, Weiß J et al (1998) Selective gold deposition via CVD onto self-assembled organic monolayers. Opt Mater 9:406–410
46. Winter C, Weckenmann U, Fischer RA et al (2000) Selective nucleation and area selective OMCVD of gold on patterend self-assembled organic monolayers studied by AFM and XPS: a comparison of OMCVD and PVD. Adv Mater CVD 6:199–205
47. Wohlfart P, Weiß J, Käshammer J et al (1999) Selective ultrathin gold deposition by organometallic chemical vapor deposition onto organic self-assembled monolayers (SAMs). Thin Solid Films 340:274–279
48. Wilchek M, Bayer EA (eds) (1990) Avidin-biotin technology, methods in enzymology, vol 184. Academic, San Diego
49. Weisser M, Tovar G, Mittler-Neher S et al (1999) Specific bio-recognition reactions observed with an integrated Mach-Zehnder interferometer. Biosens Bioelectron 14:405–411
50. Storhoff JJ, Lucas AD, Garimella V et al (2004) Homogeneous detection of unamplified genomic DNA sequences based on colorimetric scatter of gold nanoparticles probes. Nat Biol 22:883–887
51. Qi ZM, Matsuda N, Santos J et al (2004) Journal of colloid and interface science 271:249–259
52. Hamamoto K, Micheletto R, Oyama M et al (2006) An original planar multireflection system for sensing using the local surface plasmon resonance of gold nanospheres J Opt Pure Appl Opt 8:268–271

53. Aliganga AKA, Lieberwirth I, Glasser G et al (2007) Fabrication of equally oriented pancake shaped gold nanoparticles by SAM templated OMCVD and their optical response. Org Electron 8:161–174
54. Rooney P, Rezaee A, Xu S et al (2008) Control of surface plasmon resonances in dielectrically-coated proximate gold nanoparticles immobilized on a substrate. Phys Rev B 77:235446
55. Hashemi Rafsanjani SM, Rangan C, Mittler S (2008) A novel biosensing approach based on linear arrays of immobilized gold nanoparticles. Journal of Applied Physics, under revision
56. Nakanishi T, Takada H, Iida H et al (2008) Immobilization of gold nanoparticles on optical waveguides with organosilane monolayer. Colloids and Surfaces A - Physiocochemical and Engineering Aspects 331–314:234–238
57. Wang TJ, Lin CS (2006) Electro-optically modulated localized surface plasmon resonance biosensors with gold nanoparticles. Appl Phys Lett 89:173903
58. Wang TJ, Hsieh CW (2007) Phase interrogation of localized surface plasmon resonance biosensors based on electro-optic modulation. Appl Phys Lett 91:13903
59. Schmidt B, Almeida V, Manolatou C et al (2004) Nanocavity in a silicon waveguide for ultra sensitive nanoparticle detection. Appl Phys Lett 85:4854–4856
60. Hojjati H, Jiang H, Manifar T et al (2008) Photonic Crystal formation by regular pattern of gold nanoparticles within the evanescent field of a slab glass waveguide, in preparation
61. Linden L, Kuhl J, Giessen H (2001) Controlling the Interaction between light and gold nanoparticles: selective suppression of extinction. Phys Rev Lett 86:4688–4691
62. Gantzounis G, Stefanou N, Papanikolaou N (2008) Optical properties of periodic structures of metallic nanodiscs. Phys Rev B 77:035101
63. Jiang H, Manifar T, Sabarinathan J, Mittler S (2009) 3-D FDTD Analysis of gold nanoparticle based photonic crystal on slab waveguide. J Lightwave Technol 27(13):2264–2270
64. Busch K, Lölkes L, Wehrsporn RB et al (eds) (2004) Photonic crystals, advances in design, fabrication and characterization. Wiley, Weinheim
65. Meriaudeau F, Wig A, Passian A et al (2000) Gold island fiber optical sensor for refractive index sensing. Sensors Actuators B Chem 69:51057
66. Mitsui K, Handa Y, Kajikawa K (2004) Optical fiber affinity biosensor based on localized surface plasmon resonance. Appl Phys Lett 85:4231–4233
67. Sharma A, Gupta BD (2005) Fiber optic sensor based on surface plasmon resonance with nanoparticle films. Photon Nanostruct Fundamentals Appl 3:30–37
68. Sharma AK, Gupta BD (2006) Fiber-optic sensor based on surface plasmon resonance with Ag-Au alloy nanoparticle fims. Nanotechnology 17:124–131
69. Chau LK, Lin YF, Cheng SF et al (2006) Fiber-optic chemical and biochemical probes on localized surface plasmon resonance. Sensors Actuators B Chem 113:100–105
70. Tang JL, Cheng SF, Hsu WT et al (2006) Fiber-optic biochemical sensing with colloidal gold-modified long period fiber grating. Sensors Actuators B Chem 119:1050109
71. Yan H, Gu C, Yang C et al (2006) Hollow core photonic crystal fiber surface enhanced Raman probe. Appl Phys Lett 89:204101

Index

MIX
Papier aus verantwortungsvollen Quellen
Paper from responsible sources
FSC® C105338

If you have any concerns about our products,
you can contact us on
ProductSafety@springernature.com

In case Publisher is established outside the EU,
the EU authorized representative is:
Springer Nature Customer Service Center GmbH
Europaplatz 3, 69115 Heidelberg, Germany

Printed by Libri Plureos GmbH
in Hamburg, Germany